A Wealth of Numbers

A Wealth of Numbers

An Anthology

of 500 Years of Popular

Mathematics Writing

EDITED BY BENJAMIN WARDHAUGH

PRINCETON UNIVERSITY PRESS ❖ PRINCETON AND OXFORD

Copyright © 2012 by Princeton University Press

Published by Princeton University Press, 41 William Street,
Princeton, New Jersey 08540

In the United Kingdom: Princeton University Press, 6 Oxford Street,
Woodstock, Oxfordshire OX20 1TW
press.princeton.edu

Jacket Illustration: *Hardy's Taxi*, acrylic on canvas, 60 × 60 cm
© Eugen Jost, Switzerland. www.everything-is-number.net

Library of Congress Cataloging-in-Publication Data

A wealth of numbers: an anthology of 500 years of popular mathematics
writing / edited by Benjamin Wardhaugh.
p. cm.
Includes bibliographical references and index.
ISBN 978-0-691-14775-8 (hardcover : acid-free paper) 1. Mathematics.
I. Wardhaugh, Benjamin, 1979–
QA7.W43 2012
500.9—dc23   2011038672

British Library Cataloging-in-Publication Data is available

This book has been composed in Minion Pro

Printed on acid-free paper. ∞

Typeset by S R Nova Pvt Ltd, Bangalore, India

Printed in the United States of America

1 3 5 7 9 10 8 6 4 2

Contents

◦

Preface

◈

HOW DID ORDINARY PEOPLE THINK ABOUT MATHEMATICS IN THE PAST? How did they write about it? How did they learn it and teach it? If—like me—you think those questions are fascinating, read on.

Mathematics has been written about and thought about in all kinds of different ways over the centuries, and, since the beginning of printing more than 500 years ago, whole genres of mathematical writing have appeared and, often, disappeared. This book brings together a taste of many of those kinds of writing. As a result, it's more like a spice rack than a finished recipe—a rambling garden of delights rather than an orderly display of prize blooms.

That said, these hundred extracts do add up to something more than themselves: a history of mathematics which shows the subject through the eyes of the interested and the curious from the sixteenth century to the present. A history in which the changes that come are in the agendas of the writers and the interests of their readers: different mathematical audiences, different social contexts, different senses of the use of mathematics and the point of thinking about it.

So this is not a history of mathematical *research* or of new mathematics, not a story in which discoveries and innovations feature very largely or where the names of the writers are, often, ones you'll have heard before (or ever hear again).

The eleven chapters take the story in different directions, looking at how mathematics was learned and taught, used at work and played with in spare time, reflected on, and laughed about. Some chapters (1, 3, 5, and 7) look at mathematics done for fun: games and puzzles, popularizations and histories. Others (Chapters 2, 4, 6, and 8) show it in the classroom and at work. Chapters 9 and 10 are more reflective, asking *how* mathematics should be learned and taught, and *why*. And we end in Chapter 11 with my own favorite: mathematics in fiction.

The only problem with putting together this book has been an embarrassment of riches; there is just so *much* writing about mathematics aimed at ordinary people, and it is so varied in so many ways, producing a sense of almost ludicrous inadequacy in anyone who tries to make a selection. I've tried to show as much of that diversity as I can, but I hope you'll finish the book, as I do, wishing for more. It's all here: mainstream or eccentric, famous or obscure, elegant or odd. The only cutoffs are that it must be aimed at readers with school-level mathematics (or less) and that it must be published (and in English, though early translations count).

Further Reading

If you are interested in the "higher" mathematics of the past, and the mathematical writing that was written for specialists rather than ordinary people, one of the best anthologies is Jacqueline Stedall, *Mathematics Emerging: A Sourcebook 1540–1900* (Oxford, 2008). For a broad take on mathematics as a global phenomenon, there's Marcia Ascher, *Mathematics Elsewhere: An Exploration of Ideas across Cultures* (Princeton, 2002). A superlative compendium of recreational mathematics is Elwyn R. Berlekamp, John H. Conway, and Richard K. Guy, *Winning Ways for Your Mathematical Plays* (Academic Press, 1982). And, finally, if you'd like a little help in reading and thinking about mathematical writings from the past, you might try my own *How to Read Historical Mathematics* (Princeton, 2010).

Note on the Text

These extracts come from books, magazines, newspapers, and websites over a period of more than 400 years, and their original language and presentation vary extremely widely. I have striven for a consistent presentation rather than one which preserves sometimes distracting idiosyncrasies (of spelling, punctuation, or layout, for instance) for their own sake. So spelling, punctuation, and the use of italics have been modernized throughout. Obvious errors in the original texts (which are very rare) have been corrected, abbreviations expanded, and references to figures inserted, without comment. Other changes and omissions are marked with ⌐corner brackets⌐, though rather than carry this to extremes, I occasionally present a whole extract as a paraphrase where the original language seems otherwise impenetrable; these cases are noted as they appear. Longer omitted sections are marked with this symbol: ❧. The mathematical notation follows the original texts as far as possible—though modern typesetting often makes a huge difference in its appearance—and any exceptions are noted. Some extracts have a few explanatory notes, marked like this: °. They can be found at the end of the extract in question.

Acknowledgments

This book owes its existence and such merits as it possesses to the wise assistance and advice of many colleagues and friends, including Vickie Kearn, Stefani Wexler, and no fewer than six anonymous referees at Princeton University Press, as well as Jacqueline Stedall at Oxford University and, among my family, Jessica and Moira Wardhaugh. It owes its defects to me alone.

A Wealth of Numbers

∾ **1** ∾

"Sports and Pastimes, Done by Number": Mathematical Tricks, Mathematical Games

YOU'VE PROBABLY PLAYED A MATHEMATICAL GAME AT ONE TIME OR another. From the counting games we learn as children and the calculator tricks we play in the schoolyard to classics like Sprouts or Lewis Carroll's Game of Logic, there's a whole world of game playing to be had in the realm of numbers. Mathematicians used to be accused of doing magic (some still are), and while conjuring spirits or divining the future may be far from what most of us think of when we think of mathematics, there is a timeless innocent pleasure in the wool-over-the-eyes mathematical tricks of the kind that this chapter showcases.

The selections in this chapter cover the whole period from the middle of the sixteenth century to the end of the twentieth, and if they show one thing it is that tastes have not changed all that much. Some of the very first mathematics books to be printed in English contained "guess my number" tricks and questions about what happens when you double a number again and again and again: kinds that are still popular.

At the same time, there are some areas in which innovations in mathematics have opened up new ground for mathematical games and puzzles. Leonhard Euler, for instance (whom we will meet again in Chapter 7), did important work on the ways of traversing a maze or a set of paths, and this made it much easier for mathematical writers after him to set "route problems" with confidence. Rouse Ball's 1892 "recreations" included reports on this and on a mathematical problem—that of coloring a map—which was an unsolved problem in his day, and, like Alan Parr's open-ended family of Femto games in the final extract of this chapter, it shows how mathematical games can also be an invitation to explore, discover, and create for yourself.

The Well Spring of Sciences

Humfrey Baker, 1564

Humfrey Baker was a teacher in sixteenth-century London, the translator of a book on almanacs, and the author of the very successful arithmetic primer, *The Welspring of Sciences*, embodying its author's infectious enthusiasm for its subject (he once compared arithmetic to good wine, which needed no "garlande" to persuade buyers of its merits).

First published in 1562, *The Welspring* went into many editions down to 1670: the later versions were simply called *Baker's Arithmetick*. The final section of the book gave a selection of mathematical amusements, some of the first pieces of recreational mathematics to be printed in England.

Baker's dense prose is presented here in a simplified paraphrase.

Humfrey Baker (fl. 1557–1574), *The Welspring of Sciences, Which teacheth the perfecte worke and practise of Arithmetic both in vvhole numbers & fractions, with such easie and compendious instruction into the saide art, as hath not heretofore been by any set out nor laboured. Beautified vvith most necessary Rules and Questions, not onely profitable for Marchauntes, but also for all Artificers, as in the Table doth plainely appere. Novv nevvely printed and corrected. Sette forth by Humfrey Baker Citizen of London.* (London, 1564), 158^v–162^r.

If you would know the number that any man doth think or imagine in his mind, as though you could divine . . .

Bid him triple the number. Then, if the result be even, let him take half of it; if it be odd, let him take the "greater half" (that is, the next whole number above half of it). Then bid him triple again the said half. Next, tell him to cast out, if he can, 36, 27, 18, or 9 from the result: that is, ask him to subtract 9 as many times as is possible, and keep the number of times in his mind. And when he cannot take away 9 any more, tell him to take away 3, 2, or 1, if he can, so as to find out if there is anything left besides the nines.

This done, ask how many times he subtracted 9. Multiply this by 2. And if he had any thing remaining beside the nines, add 1.

For example, suppose that he thought of 6. Being tripled it is 18, of which a half is 9. The triple of that is 27; now ask him to subtract 18, or 9, or 27, and again 9. But then he will say to you that he cannot; ask him to subtract 3, or 2, or 1. He will say also that he cannot; thus, considering that you have made him to subtract three times 9, you shall tell him that he thought of 6, for 3 times 2 makes 6.

If he had thought of 5, the triple of it is 15, of which the "greater half" is 8. The triple of that makes 24, which contains two nines. Two times two makes four, and since there is something remaining we add 1. This makes 5, which is the number that he thought of.

If someone in a group has a ring upon his finger, and you wish to know, as though by magic, who has it, and on which finger and which joint . . .

Ask the group to sit down in order, numbering themselves 1, 2, 3, etc. Then leave the room, and ask one of the onlookers to do the following. Double the number of the person that has the ring, and add 5. Then multiply by 5, and add the number of the finger on which the ring is. Then ask him to append to the result the figure (1, 2, or 3) signifying which joint the ring is on. (Suppose the result was 89 and the ring was on the third joint; then he will make 893.)

This done, ask him what number he has. From this, subtract 250, and you will have a number with at least three digits. The first will be the number of the person who has the ring. The second will be the number of the finger. And the last will be the number of the joint. So, if the number was 893, you subtract 250, and there will remain 643. Which shows you that the sixth person has the ring on the fourth finger, and on the third joint.

But note that when you have made your subtraction, if there is a zero in the tens—that is, in the second digit—you must take one from the hundreds digit. And that "one" will be worth ten tenths, signifying the tenth finger. So, if there remains 703, you must say that the sixth person has the ring on his tenth finger and on the third joint.

In the same way, if a man casts three dice, you may know the score of each of them.

Ask him to double the score of one die, add 5, and then multiply by 5. Next, add the score of one of the other dice, and append to the result the score of the last die. Then ask him what number he has. Subtract 250, and there will remain 3 digits, which tell you the points of the three dice.

Similarly, if three of your companions—say, Peter, James, and John—give themselves different names in your absence—for example, Peter would be called a king, James a duke, and John a knight—you can divine which of them is called a king, which a duke, and which a knight.

Take twenty-four stones (or any other tokens), and, first, give one to one of your friends. Next, give two to another of them, and finally give three to the last of them. Keep a note of the order in which you have given them the stones. Then, leaving the eighteen remaining stones before them, leave the room or turn your back, saying: "whoever calls himself a king, for every stone that I gave him let him take one of the remaining ones; he that calls himself a duke, for every stone that I gave him let him take two of them that remain; and he that calls himself a knight, for every stone that I gave him let him take four."

This being done, return to them, and count how many stones are left. There cannot remain any number except one of these: 1, 2, 3, 5, 6, 7. And for each of these we have chosen a special name, thus: *Angeli, Beati, Qualiter, Messias, Israel, Pietas*. Each name contains the three vowels *a, e, i*, and these show you the names in order. *A* shows which is the king, *E* which is the duke, and *I* shows which is the knight, in the same order in which you gave them the stones. Thus, if there remains only one stone, the first name, *Angeli*, shows by the vowels *a, e, i* that your first friend is the king, the second the duke, and the third the knight. If there remain two stones, the second name, *Beati*, shows you by the vowels *e, a, i* that your first friend is the duke, the second the King, and the third the knight. And so on for the other numbers and names.

Mathematical Recreations

Henry van Etten, 1633

Henry van Etten's *Mathematicall Recreations*, first published in French in 1624, collected together a wide variety of different material. Some of the "problems" were physical tricks or illusions, like "How a Millstone or other ponderosity may hang upon the point of a Needle without bowing, or any wise breaking of it." Others were numerical tricks like those in Baker's *Welspring of Sciences*, above, and still others were optical effects or illusions. The extracts given below thus show some of the diversity in what could plausibly be called mathematics at the time: a diversity which is emphasized by the book's splendidly encyclopedic title. They include a remarkable early report of what Galileo had seen through his telescope, together with the cheery assertion that making a good telescope was a matter of luck ("hazard") as much as skill.

Van Etten was apparently a pseudonym of the French Jesuit Jean Leurechon (c. 1591–1670). The translation has been ascribed to various different people, but its real author remains a mystery.

Henry van Etten (trans. anon.), *Mathematicall Recreations. Or a Collection of sundrie Problemes, extracted out of the Ancient and Moderne Philosophers, as secrets in nature, and experiments in Arithmetic, Geometrie, Cosmographie, Horologographie, Astronomie, Navigation, Musicke, Opticks, Architecture, Staticke, Machanicks, Chimestrie, Waterworkes, Fireworks, etc. Not vulgarly made manifest untill this time: Fit for Schollers, Students, and Gentlemen, that desire to knovv the Philosophicall cause of many admirable Conclusions. Vsefull for others, to acuate and stirre them up to the search of further knowledge; and serviceable to all for many excellent things, both for pleasure and Recreation. Most of which were written first in Greeke and Latine, lately compiled in French, by Henry Van Etten Gent. And now delivered in the English tongue, with the Examinations, Corrections and Augmentations.* (London, 1633), pp. 47–50, 98–102, 167, 208–209, 240.

How to describe a Circle that shall touch 3 Points placed howsoever upon a plane, if they be not in a ˌstraightˌ line

Let the three points be A, B, C. Put one foot of the Compass upon A and describe an Arc of a Circle at pleasure; and placed at B, cross that Arc in the two points E and F; and placed in C, cross the Arc in G and H. Then lay a ruler upon GH and draw a line, and placˌingˌ a Ruler upon E and F, cut the other line in K. So K is the Center of the Circumference of a Circle, which will pass by the said three points A, B, C.

Or it may be inverted: having a Circle drawn, to find the Center of that Circle. Make 3 points in the circumference, and then use the same way: so shall you have the Center, a thing most facile to every practitioner in the principles of Geometry.

How to change a Circle into a square form

Make a Circle upon pasteboard or other material, and ⌊label the centre A⌋; then cut it into 4 quarters, and dispose them so that A, at the center of the Circle, may always be at the Angle of the square. And so the four quarters of the Circle being placed so, it will make a perfect square, whose side AA is equall to the diameter. Now here is to be noted that the square is greater than the Circle by the vacuity in the middle.

With one and the same compasses, and at one and the same extent, or opening, how to describe many Circles concentrical, that is, greater or lesser one than another

It is not without cause that many admire how this proposition is to be resolved; yea, in the judgement of some it is thought impossible, who consider not the industry of an ingenious Geometrician, who makes it possible: and that most facile, sundry ways. For in the first place, if you make a Circle upon a fine plane, and upon the Center of that Circle a small peg of wood be placed, to be raised up and put down at pleasure by help of a small hole made in the Center, then with the same opening of the Compasses you may describe Circles Concentrical: that is, one greater or lesser than another. For the higher the Center is lifted up, the lesser the Circle will be.

Secondly, the compass being at that extent upon a Gibbous body, a Circle may be described, which will be less than the former, upon a plane, and more artificially upon a Globe, or round bowl. And this again is most obvious upon a round Pyramid, placing the Compasses upon the top of it, which will be far less than any of the former; and this is demonstrated by the 20⌊th⌋ Pro⌊position⌋ of the first ⌊book⌋ of Euclid's ⌊*Elements*⌋.

Of spectacles of pleasure

⌊...⌋ Now I would not pass this Problem without saying something of Galileo's admirable Glass: for the common simple perspective Glasses give to aged men but the eyes or sight of young men, but this of Galileo gives a man an Eagle's eye, or an eye that pierceth the heavens. First it discovereth the spotty and shadowed opacous bodies that are found about the Sun, which darkeneth and diminisheth the splendour of that beautiful and shining Luminary; secondly, it shows the new planets that accompany Saturn and Jupiter; thirdly, in Venus is seen the new, full, and quartal increase, as in the Moon by her separation from the Sun; fourthly, the artificial structure of this instrument helpeth us to see an innumerable number of stars, which otherwise are obscured by reason of the natural weakness of our sight. Yea, the stars in ⸢the Milky Way⸣ are seen most apparently; where there seems no stars to be, this instrument makes apparently to be seen, and further delivers them to the eye in their true and lively colour, as they are in the heavens: in which the splendour of some is as the Sun in his most glorious beauty. This Glass hath also a most excellent use in observing the body of the Moon in time of Eclipses, for it augments it manifold, and most manifestly shows the true form of the cloudy substance in the Sun, and by it is seen when the shadow of the Earth begins to eclipse the Moon, and when totally she is overshadowed.

Besides the celestial uses which are made of this Glass, it hath another notable property: it far exceedeth the ordinary perspective Glasses which are used to see things remote upon the Earth, for as this Glass reacheth up to the heavens and excelleth them there in his performance, so on the Earth it claimeth preeminency. For the objects which are farthest remote, and most obscure, are seen plainer than those which are near at hand, scorning, as it were, all small and trivial services, as leaving them to an inferior help. Great use may be made of this Glass in discovering of Ships, Armies, etc.

Now the apparel or parts of this instrument or Glass is very mean or simple, which makes it the more admirable (seeing it performs such great service), having but a convex Glass, thickest in the middle, to unite and amass the rays, and make the object the greater ⸢...⸣ augmenting the visual Angle. As also a pipe or trunk to amass the Species, and hinder the greatness of the light which is about it (to see well, the object must be well enlightened, and the eye in obscurity). Then there is adjoined unto it a Glass of a short sight to distinguish the rays, which the other would make more confused if alone. As for the proportion of those Glasses to the Trunk, though there be certain rules to make them, yet it is often by hazard that there is made an excellent one, there being so many difficulties in the action, therefore many ought to be tried, seeing that exact proportion, in Geometrical calculation, cannot serve for diversity of sights in the observation.

Of the Dial upon the fingers and the hand

Is it not a commodity very agreeable, when one is in the field or in some village without any other Dial, to see only by the hand what of the clock it is? which gives it very near, and may be practised by the left hand in this manner.

Take a straw, or like thing, of the length of the Index or the second finger. Hold this straw very tight between the thumb and the right finger, then stretch forth the hand and turn your back and the palm of your hand towards the Sun, so that the shadow of the muscle which is under the thumb touch,es, the line of life, which is between the middle of the two other great lines, which is seen in the palm of the hand. This done, the end of the shadow will show what of the clock it is: for at the end of the great finger it is 7 in the morning or 3 in the evening; at the end of the Ring finger it is 8 in the morning or 4 in the evening; at the end of the little finger or first joint, it is 9 in the morning or 3 in the afternoon; 10 and 2 at the second joint; 11 and 1 at the third joint; and midday in the line following, which comes from the end of the Index ,finger,.

Of sundry Questions of Arithmetic, and first of the number of sands

It may be said, incontinent,ly,, that to undertake this were impossible, either to number the sands of Libya, or the sands of the Sea. And it was this that the Poets sung, and that which the vulgar believes—nay, that which long ago certaine Philosophers to Gelon King of Sicily reported: that the grains of sand were innumerable. But I answer, with Archimedes,° that not only one may number those which are at the border and about the Sea, but those which are able to fill the whole world, if there were nothing else but sand, and the grains of sands admitted to be so small that 10 may make but one grain of Poppy. For at the end of the account there need,s, not,hing, to express them but this number: 30,840,979,456, and 35 Ciphers at the end of it. Clavius and Archimedes make it somewhat more, because they make a greater firmament than Tycho Brahe doth, and if they augment the Universe, it is easy for us to augment the number, and declare assuredly how many grains of sand there is requisite to fill another world, in comparison that our visible world were but as one grain of sand, an atom or a point. For there is nothing to do but to multiply the number by itself, which will amount to ninety places, whereof twenty are these: 95,143,798,134,910,955,936, and 70 Ciphers at the end of it. Which amounts to a most prodigious number ⌞...⌟.

To measure an inaccessible distance, as the breadth of a River, with the help of one's hat only

The way of this is easy: for having one's hat upon his head, come near to the bank of the River, and holding your head upright (which may be by putting a small stick to some one of your buttons to prop up the chin), pluck down the brim or edge of your hat until you may but see the other side of the water. Then turn about ⌞...⌟ in the same posture ⌞as⌟ before, towards some plain, and mark where the sight, by the brim of the hat, glanceth on the ground: for the distance from that place to your standing, is the breadth of the River required.

Note

Archimedes (c. 287–c. 212 BC) had attempted to calculate the number of grains of sand the universe could contain in his *Sand Reckoner*, reaching a result of 8×10^{63}.

"How Prodigiously Numbers Do Increase"

William Leybourne, 1667

William Leybourne wrote on a range of subjects including astronomy, geography, and surveying, reflecting a career which took in a period as a bookseller and printer, as well as his later roles of mathematician, teacher, and surveyor.

His book of "recreations" begins with parlor tricks and ends with a set of strategies to help with arithmetic; some of the impressive wealth of material was in fact taken from Henry van Etten (see the previous extract). One section, shown here, contains several variations on the "geometric progression" story that, today, is sometimes told of a grain of rice doubled for each of the squares on a chessboard. Leybourne evidently felt that his readers might have difficulty reading the very large numbers involved and took pains to write them out both in figures and in words.

William Leybourne (1626–1716), *Arithmetical Recreations: Or, Enchiridion of Arithmetical Questions: Both Delightful and Profitable. Whereunto are added Diverse Compendious Rules in Arithmetic, by which some seeming difficulties are removed, and the performance of them rendred familiar and easie to such as desire to be Proficients in the Science of Numbers. All performed without Algebra. By Will. Leybourne.* (London, 1667), pp. 122–140.

Concerning two Neighbours Changing of their Land

Two Neighbours had either of them a piece of Land: the one field was four-square, every side containing 120 perches,° so that it was round about 480 perches; the other was square also, but the sides longer than the other's field, and the ends shorter, for the sides of this field were 140 perches long apiece, and the ends thereof were 100 perches apiece, so that this field was 480 perches about as well as the other. Now, which of these two had the best bargain?

It is wonderful to see how Numbers will discover that to be erroneous and absurd, which to common sense and man's apprehension appears reasonable, as in this bargain. First, for the field which is 120 perches on either side: multiply 120 by 120, and the product is 14,400, and so many square perches doth that piece of Land contain, that is, 80 acres. Then, for the other field: multiply 140 (the length of one of the longer sides of the field) by 100 perches (one of the shorter sides) and the product will be 14,000 ⌊square⌋ perches, the content of the other field, which is less then the former by 400 ⌊square⌋ perches, that is, 2 acres and an half. And so much would he have lost that had the field of 120 perches on every side, though the other field were as much about.

And this error would still grow greater, the narrower the second field had been. As, suppose the ends or shorter sides thereof had been but 40 perches apiece, and the longer sides 200 apiece; this field would still have been 480 perches about, but let us see how much it contains. Multiply 200, the longer side, by 40, the shorter side, and it will produce 8000, and so many square perches will it contain, which is but 50 acres. So that if he had changed for this field, which is as much about as his, he would have then lost 30 acres by the bargain.

About the borrowing of Corn

A Country Farmer had in his house a vessel of Wood full of Wheat, which was 4 foot high, 4 foot broad both at top and bottom, and in all parts 4 foot, as the sides of a Die. One of his Neighbours desires him to lend him half his Wheat till Harvest, which he doth. Harvest coming, and his Neighbour is to repay, he makes a Vessel 2 foot every way, and fills him that twice, in lieu of what he borrowed. Was there gain or loss in this particular?

Examine first what either of these Vessels will hold, and by that you will discover the fallacy. First, the vessel 4 foot high contains 48 inches of a side, wherefore multiply 48 by 48, and the product will be 2304, which multiply again by 48, and the product will be 110,592; and so many cubical or square inches of Corn do his vessel hold, the half whereof, which is 55,296, he lent his Neighbour. Now, secondly, let us examine how much the second vessel will hold, it being 2 foot on every side, that is 24 inches. Multiply 24 by 24; the product is 574. Which multiply again by 24, and the product will be 13,824; and so many cubical or square inches did the lesser vessel contain. Which being filled twice, it made 27,648 cubical inches of corn or wheat, which was all he paid his Neighbour in lieu of the 55,296 inches which he borrowed, which is but the just half. And so allowing 2256 cubical inches of Wheat to make a Bushel (for so many there is in a Bushel) he paid his Neighbour less by 12 bushels, and about a peck, than he borrowed of him. And this, and the reason of it, is evident, as I will demonstrate to you by a familiar precedent. If you cause a Die to be made of one inch every side, and 8 other Dice to be made of half an inch every side, these 8 being laid close, one to another, in a square form, these 8 will be but of the same bigness with the other one Die, whose side is but an inch.

A Bargain between a Farmer and a Goldsmith

A rich Farmer being in a Fair, espies at a Goldsmith's Shop a Necklace of Pearl, upon which were 72 Pearls. The Farmer cheapening of it, the Goldsmith asked 30 shillings a Pearl, at which rate the Necklace would come to £103. The Farmer, looking upon it as dear, goes his way, offering nothing. Whereupon the Goldsmith calls him, and tells him, if he thought much to part with Money, he would deal with him for Corn. To which the Farmer hearkens, asks him how much Corn he would have for it at two shillings the Bushel. The Goldsmith told him he would be very reasonable, and would take for the first Pearl one Barley corn only, for the second two corns, for the third four corns, and so doubling the corns till the 72 Pearls were out. To this the Farmer agrees, and immediately strikes the Bargain. But see the event.

He that hath any skill in Numbers, will easily discern the vanity that there is in this kind of bargaining, so that no man can be bound to them; for Numbers increasing in a Geometrical Progression do so prodigiously increase, that (to those that are ignorant of the reason) it will seem impossible they should do so. But that it is so will appear evident by this bargain, if you enquire: first, the quantity; secondly, the worth of so much Barley in Money; and thirdly, the weight of it, and how it should be removed, or where stowed. Wherefore,

1. If we allow 10,000—ten thousand—corns to a Pint (which is more than enough) then 5,120,000 corns will make a Quarter, but yet (for the ease of them that will make tryal) we will allow 10,000,000 corns to make a Quarter. By which number, if you divide the whole number of corns that the 72 Pearls would have amounted unto ⌞...⌟ the Quotient of that Division would be

472,236,648,286,964.

And so many whole Quarters of Barley would the Necklace have amounted unto, and some odd Bushels, which we here omit as superfluous.

2. Now for the worth of this Barley, suppose it were sold at 13 pence the Bushel (which is a reasonable rate), that is, 10 shillings the Quarter. Wherefore, divide the foregoing number of Quarters by 2—that is, take half of it—and it will be 236,118,324,143,482 pounds sterling; which sum rendered in words, is, *Two hundred thirty-six millions of millions, one hundred and eighteen thousand, three hundred twenty-four millions, one hundred forty-three thousand, four hundred eighty-two* pounds. A vast sum of money for a Farmer. ⌞...⌟ So great vanity may be agreed and concluded upon by people ignorant of this Science, and for want of serious premeditation. But,

3. Let us consider the weight of so much Barley. If we allow 8 Bushels (or one Quarter) to weigh Two hundredweight (but doubtless it weighs more), then the whole number of Quarters, multiplied by 2, gives the weight of all the Barley to be 944,473,296,573,928 hundredweight. And if you divide this number by 20, the Quotient will be 47,223,664,828,696 Tons; that is, *Forty-seven millions of millions, two hundred twenty-three thousand six hundred sixty-four millions, eight hundred twenty-eight thousand, six hundred ninety-six* Tons. Which will require 47,223,664,828—that is, *Forty-seven thousand two hundred twenty-three millions, six hundred*

sixty-four thousand, eight hundred twenty-eight—Ships of a thousand Ton apiece to carry it. And to conclude, If there were four Millions of Nations in the World, and every one of those Nations had Ten thousand Sail of such Ships of a Thousand Ton apiece, yet all those Ships would not contain it. Thus, by this, you may see how prodigiously numbers do increase, being multiplied according to Geometrical Progression.

Concerning an Agreement that a Country-Fellow made with a Farmer

A Country-Fellow comes to a Farmer, and offers to serve him for 8 years, all which time he would require no other Wages than One grain of Corn, and one quarter of an inch of Land to sow it in the first year, and Land enough to sow that one Corn, and the increase of it, for his whole 8 years: to which the Farmer assents.

Their Bargain being thus made, let us consider what his eight years' service will be worth. For the first year he hath only one quarter of an inch of Ground, and one Corn, which Corn we will suppose had in the Ear at the year's end 40 Corns (for that is few enough). Then the second year he must have 40 square quarters of inches of ground to sow those 40 Corns in: that is, 10 square inches of ground. And the third year, supposing those 40 Corns to produce 40 Ears, and in each Ear 40 Corns, as before, they will be in the third year increased to 1600 corns, so that he must have 1600 square quarters of inches to sow that increase in, which is 40 square inches. And thus continuing till the 8 years be expired, the increase would be 6,553,600,000,000 corns—that is, *Six millions of millions, five hundred fifty-three thousand and six hundred millions*—of corns, and so many square quarters of inches of Land must he have to sow this increase in. Now know that 3,600,000,000—that is, *Three thousand and six hundred millions*—of square inches do make a mile, upon the ⸤surface⸥ or plane, and that ⸤a square⸥ Mile will be capable to receive 14,400,000,000—that is, *Fourteen thousand and four hundred millions*—of Corns. Wherefore divide 6,553,600,000,000 (the whole number of the Corn's increase) by 14,400,000,000 (the number of Corns that one Mile square of Land is capable to receive), and in the Quotient you shall have 455; and so many miles square of Land must there be, to contain the sowing of the increase of one Corn in 8 years. Which will be about 420,000—that is *Four hundred and twenty thousand*—Acres of Land, which being rated at half a Crown an Acre by the year, it will amount unto 50,000—that is, *Fifty thousand*—pound, which is 6,250—that is, *Six thousand two hundred and fifty*—pound a year: a very considerable Salary for Eight years' service.

By these, and the like Contract.ₛ., we may see what absurdities are and may easily be committed, which you see the subtlety of Numbers easily discovers. I will give you, by way of Discourse, some precedents concerning the increase of Creatures of several kinds, which you may make trial of for your Recreation. As,

1. What think you, if one should tell the Great Turk (or any other Potentate in the World, who having a great Army in the Field in continual pay) that all the Revenue appertaining to his Crown, will not for one year's time maintain all the Pigs that one Sow, with all the Pigs of her race, and the increase issuing of them, shall produce in 12 years, notwithstanding he can maintain so great an Army in the field? Doubtless, he would take it unkindly. But consider: imagine the Sow brings forth but 6 Pigs at a Litter, of which we will allow 2 to be Barrow (and this supposition is as little as may be), and then imagine that every of those 4 bring as many every year, and the increase of them the like, during the term of 12 years. They and their race at the end of the time will be increased to *Three and thirty millions* of Pigs. Now if we allow 5 shillings for to maintain one of these Pigs for a year, which is not full half a farthing a day, yet there must be *Three and thirty millions* of Crowns to maintain them one year: which will make a great hole in a large Revenue.

2. Would any man (that hath not skill in Numbers) imagine that 100 Sheep, and the increase thereof, being preserved for the space of 16 years, should be worth above *One million, six hundred and twenty thousand* pounds sterling? Yet if every Sheep do produce but one every year, at the expiration of 16 years, 100 sheep will increase unto 6,553,600, which is *Six millions, five hundred fifty-three thousand, six hundred* Sheep. Now supposing these to be worth but 5 shillings apiece, they would at that rate amount unto 1,638,400—that is, *One million, six hundred thirty-eight thousand four hundred*—pounds sterling.

Note

perches: measure of length, eventually standardized as $5\frac{1}{2}$ yards, or $16\frac{1}{2}$ feet, about 5.03 meters.

Profitable and Delightful Problems

Jacques Ozanam, 1708

Jacques Ozanam, "Professor of the mathematicks at Paris," published many mathematical works, including mathematical tables and a version of the *Arithmetic* of Diophantus, from the 3rd century AD. His *Recreations* first appeared in French in 1694; the mathematical parts of the book presented a range of mathematical puzzles and tricks, geometrical constructions, and other diversions broadly similar to those of his predecessors like van Etten and Leybourn. The "physical" recreations featured a memorable section on pyrotechnics which began with "Problem I. To make Gun-Powder" and ended with "Problem XXXVIII. To make an Ointment excellent for Curing all sorts of Burnings."

Jacques Ozanam (1640–1717), *Recreations mathematical and Physical; Laying down, and Solving Many Profitable and Delightful Problems of Arithmetic, Geometry, Opticks, Gnomonicks, Cosmography, Mechanics, Physicks, and Pyrotechny* (London, 1708), pp. 82–83, 192, 350–351.

After filling one Vessel with Eight Pints of any Liquor, to put just one half of that Quantity into another Vessel that holds Five Pints, by means of a third Vessel that will hold three Pints

This Question is commonly put after the following manner: A certain Person, having a Bottle filled with 8 Pints of excellent Wine, has a mind to make a Present of the Half of it, or 4 Pints, to one of his Friends. But he has nothing to measure it out with but two other Bottles, one of which contains 5, and the other 3 Pints. ⌐I ask⌐, how he shall do to accomplish it?

To answer this Question, let's call the Bottle of 8 Pints *A*, the 5-Pint Bottle *B*, and the 3-Pint Bottle *C*. We suppose there are 8 Pints of Wine in the Bottle *A*, and the other two, *B* and *C*, are empty, as you see in ⌐row⌐ *D* (see the table below). Having filled the Bottle *B* with Wine out of the Bottle *A*, in which there will then remain but 3 Pints, as you see at *E*, fill the Bottle *C* with Wine out of the Bottle *B*, in which, by consequence, there will then remain but 2 Pints, as you see at *F*. This done, pour the Wine of the Bottle

	Bottle		
	A (8)	B (5)	C (3)
D	8	0	0
E	3	5	0
F	3	2	3
G	6	2	0
H	6	0	2
I	1	5	2
K	1	4	3

C into the Bottle A, where there will then be 6 Pints, as you see in G, and pour the 2 Pints of the Bottle B into the Bottle C, which will then have 2 Pints, as you see at H; then fill the Bottle B with Wine out of the Bottle A, by which means there will remain but 1 Pint in it, as you see at I, and conclude the Operation by filling the Bottle C with Wine out of the Bottle B, in which there will then remain just 4 Pints, as you see at K, and so the Question is solved.

To measure an accessible Line upon the Ground by means of the Flash and the Report of a Cannon

With a Musket Ball, make a Pendulum 11 Inches and 4 Lines long, calculating the Length from the Center of the Motion to the Center of the Ball; and the very moment that you perceive the Flash of the Cannon (which must be at the very place, the Distance of which, from the place where you are, is enquired after) put the Pendulum in motion, so that the Arches of the Vibrations do not exceed 30 Degrees. Multiply by 200 the Number of the Vibrations from the moment you perceived the Flash to the moment in which you hear the Report, and reckon as many Paris *Toises*° for the Distance of the place where the Gun was fired, from the place where you stood.

   Much after the same manner you may measure the Height of a Cloud, when 'tis near the Zenith, and at a time of Thunder and Lightning. But this way of measuring Distances is very uncertain, and I only mentioned it here as a Recreation.

To make a deceitful Balance, that shall appear just and even both when empty and when loaded with unequal Weights

Make a Balance, the Scales of which are of unequal Weight, and of which the two Arms are of unequal length, and in reciprocal proportion to these unequal Weights. That is, the first weight is to the second, as the second length is to the first: for thus the two scales will continue *in Æquilibrio* round the fixed point. And the same will be the Case if the two Arms are of equal length and of unequal thickness, so that the thickness of the first is to that of the second as the weight of the second scale is to that of the first. This supposed, if you put into the two scales unequal weights which have the same Ratio with the Gravities of the two scales, the heavier weight being in the heavier scale, and the lighter in the lighter scale, these two Weights and Scales will rest *in Æquilibrio*.

We'll suppose that the first Arm is three ⌊Inches⌋, and the second Arm two ⌊Inches⌋, and reciprocally the second scale weighs three Ounces, and the first scale two; in which case the balance will be even when they are empty. Then we put a weight of two pound into the first scale, and one of three into the second, or else one of four into the first, and one of six into the second, etc. And the balance continues still even, because the weights with the gravity of the Scales are reciprocally proportional to the length of the Arms of the Beam. Such a pair of Scales is discovered by shifting the weights from one side to another, for then the Balance will cast to one side.

Note

Toise: 6 feet.

Lotteries and Mountebanks

L. Despiau, 1801

Although it contains some of the usual card tricks, number games, and basic arithmetic (see van Etten and Leybourne, above), a fair part of Despiau's book is taken up with the attempt to make mathematics amusing by applying it to dice games. Despiau (I have failed to discover either his dates or his first name) had been a professor of mathematics and philosophy in Paris, and his book was commended by Charles Hutton, who taught at the Royal Military Academy at Woolwich. Yet it seems to have occurred to neither of them that this thematic angle, and the attached "large table of the chances or odds at play" might make the book less than perfectly suited for use in schools.

L. Despiau, *Select Amusements in Philosophy and Mathematics; Proper for agreeably exercising the Minds of Youth. Translated from the French* ⌞...⌟ *with several corrections and additions, particularly a large table of the chances or odds at play. The whole recommended as an useful Book for Schools, by Dr. Hutton, Professor of Mathematics, at Woolwich.* (London, 1801), pp. 75–77, 94–95.

Application of the doctrine of combinations to games of chance and to probabilities

Though nothing, on the first view, seems more foreign to the province of the mathematics, than games of chance, the powers of analysis have, as we may say, enchained this Proteus, and subjected it to calculation: it has found means to measure the different degrees of the probability of certain events, and this has given rise to a new branch of mathematics, the principles of which we will here explain.

When an event can take place in several different ways, the probability of its happening in a certain determinate manner is greater when, in the whole of the ways in which it can happen, the greater number of them determine it to happen in that manner. In a lottery, for example, everyone knows that the probability of obtaining a prize is greater, according as the number of the prizes is greater, and as the whole number of the tickets is less. The probability of an event, therefore, is ⌞proportional to⌟ the number of cases in which it can happen, taken directly, and ⌞to⌟ the total number of those in which it can be varied, taken inversely; consequently it may be expressed by a fraction, having for its numerator the number of the favourable chances, and for its denominator the whole of the chances.

Thus, in a lottery containing 1000 tickets, 25 of which only are prizes, the probability of obtaining a prize will be represented by $\frac{25}{1000}$ or $\frac{1}{40}$; if there were 50 prizes, the probability would be double, for in that case it would be equal to $\frac{1}{20}$; but if the number of tickets, instead of 1000, were 2000, the probability would be only one half of the former, or $\frac{1}{80}$; and if the whole number of the tickets were infinitely great, the prizes remaining the same, it would be infinitely small, or 0; while, on the other hand, it would become certainty, and be expressed by unity, if the number of the prizes were equal to that of the tickets.

Another principle of this theory, the truth of which may be readily perceived, and which it is necessary here to explain, is as follows:

A person plays an equal game when the money staked, or risked, is in the direct ratio of the probability of winning; for to play an equal game is nothing else than to deposit a sum so proportioned to the probability of winning, that, after a great many throws, the player may find himself nearly at par; but for this purpose, the stakes must be proportioned to the probability which each of the players has in his favour. Let us suppose, for example, that A bets against B on a throw of the dice, and that there are two chances in favour of the former, and one for the latter; the game will be equal if, after a great number of throws, they separate nearly without any loss. But as there are two chances in favour of A, and only one for B, after 300 throws A will have won nearly 200, and B 100. A, therefore, ought to deposit 2 and B only 1, for by these means A, in winning 200 throws, will get 200, and B, in winning 100, will get 200 also. In such cases, therefore, it is said that there is two to one in favour of A.

⌊The mountebank⌋

A mountebank, at a country fair, amused the populace with the following game: he had 6 dice, each of which was marked only on one face, the first with 1, the second with 2, and so on to the sixth, which was marked 6; the person who played gave him a certain sum of money, and he engaged to return it a hundredfold if, in throwing these six dice, the six marked faces should come up only once in 20 throws.

Though the proposal of the mountebank does not, on the first view, appear very disadvantageous to those who entrusted him with their money, it is certain that there were a great many chances against them.

It may indeed be seen that, of the 46,656 combinations of the faces of 6 dice, there is only one which gives the 6 marked faces uppermost; the probability therefore of throwing them, at one throw, is expressed by $\frac{1}{46656}$: and as the adventurer was allowed 20 throws, the probability of his succeeding was only $\frac{20}{46656}$, which is nearly equal to $\frac{1}{2332}$. To play an equal game, therefore, the mountebank would have engaged to return 2332 times the money deposited.

Dodging the Mastodon and the Plesiosaurus

Henry Ernest Dudeney, 1917

Henry Ernest Dudeney's well-loved classic, *Amusements in Mathematics*, takes quite a spare approach, giving no solutions and no mathematical analysis of its puzzles. The section on route puzzles presented here, for instance, could—and would, in the hands of many later popularizers—have been a springboard for a general discussion of unicursal problems and their mathematics, but Dudeney gives us the problems alone, often charmingly illustrated and whimsically described. If they "enable one to generalise," as he claims, it is up to the reader to work out how. In this respect, perhaps, the book recalls some of the very earliest presentations of mathematical recreations we have seen, like Baker's in the sixteenth century.

Henry Ernest Dudeney (1857–1930), *Amusements in Mathematics* (London, 1917), pp. 68–71.

Unicursal and Route Problems

It is reasonable to suppose that from the earliest ages one man has asked another such questions as these: "Which is the nearest way home?" "Which is the easiest or pleasantest way?" "How can we find a way that will enable us to dodge the mastodon and the plesiosaurus?" "How can we get there without ever crossing the track of the enemy?" All these are elementary route problems, and they can be turned into good puzzles by the introduction of some conditions that complicate matters. A variety of such complications will be found in the following examples. I have also included some enumerations of more or less difficulty. These afford excellent practice for the reasoning faculties, and enable one to generalize in the case of symmetrical forms in a manner that is most instructive.

A Juvenile Puzzle

For years I have been perpetually consulted by my juvenile friends about this little puzzle. Most children seem to know it, and yet, curiously enough,

Figure 1.1. A juvenile puzzle, with two juveniles. (Dudeney, p. 69.)

they are invariably unacquainted with the answer. The question they always ask is, "Do, please, tell me whether it is really possible." I believe Houdini the conjurer used to be very fond of giving it to his child friends, but I cannot say whether he invented the little puzzle or not. No doubt a large number of my readers will be glad to have the mystery of the solution cleared up, so I make no apology for introducing this old "teaser."

The puzzle is to draw with three strokes of the pencil the diagram that the little girl is exhibiting in the illustration (Figure 1.1). Of course, you must not remove your pencil from the paper during a stroke or go over the same line a second time. You will find that you can get in a good deal of the figure with one continuous stroke, but it will always appear as if four strokes are necessary.

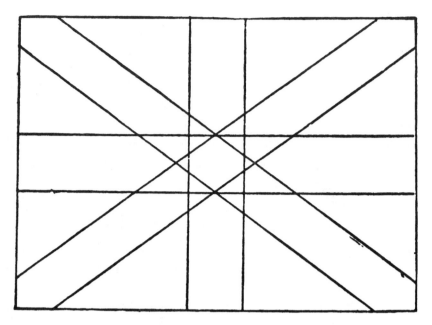

Figure 1.2. A rough sketch somewhat resembling the British flag. (Dudeney, p. 69.)

Another form of the puzzle is to draw the diagram on a slate and then rub it out in three rubs.

The Union Jack

The illustration (Figure 1.2) is a rough sketch somewhat resembling the British flag, the Union Jack. It is not possible to draw the whole of it without lifting the pencil from the paper or going over the same line twice. The puzzle is to find out just how much of the drawing it is possible to make without lifting your pencil or going twice over the same line. Take your pencil and see what is the best you can do.

The Dissected Circle

How many continuous strokes, without lifting your pencil from the paper, do you require to draw the design shown in our illustration (Figure 1.3)? Directly you change the direction of your pencil it begins a new stroke. You

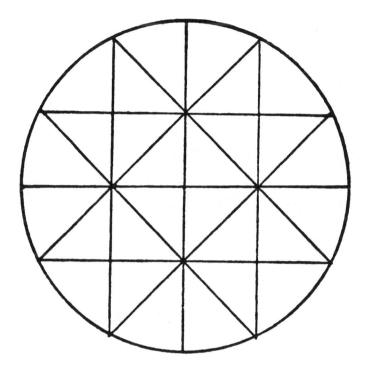

Figure 1.3. A dissected circle. (Dudeney, p. 69.)

may go over the same line more than once if you like. It requires just a little care, or you may find yourself beaten by one stroke.

The Tube Inspector's Puzzle

The man in our illustration (Figure 1.4) is in a little dilemma. He has just been appointed inspector of a certain system of tube railways, and it is his duty to inspect regularly, within a stated period, all the company's seventeen lines connecting twelve stations, as shown on the big poster plan that he is contemplating. Now he wants to arrange his route so that it shall take him over all the lines with as little travelling as possible. He may begin where he likes and end where he likes. What is his shortest route?

Could anything be simpler? But the reader will soon find that, however he decides to proceed, the inspector must go over some of the lines more than once. In other words, if we say that the stations are a mile apart, he will have to travel more than seventeen miles to inspect every line.

Figure 1.4. The tube inspector in a little dilemma. (Dudeney, p. 69.)

There is the little difficulty. How far is he compelled to travel, and which route do you recommend?

Visiting the Towns

A traveller, starting from town No. 1 (see Figure 1.5), wishes to visit every one of the towns once, and once only, going only by roads indicated by straight lines. How many different routes are there from which he can

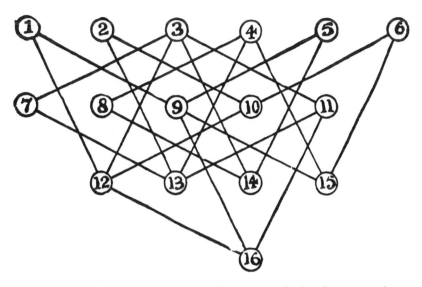

Figure 1.5. Visiting the towns, an absurdly easy puzzle. (Dudeney, p. 69.)

select? Of course, he must end his journey at No. 1, from which he started, and must take no notice of cross roads, but go straight from town to town. This is an absurdly easy puzzle, if you go the right way to work.

The Fifteen Turnings

Here is another queer travelling puzzle, the solution of which calls for ingenuity. In this case the traveller starts from the black town (see Figure 1.6) and wishes to go as far as possible while making only fifteen turnings and never going along the same road twice. The towns are supposed to be a mile apart. Supposing, for example, that he went straight to A, then straight to B, then to C, D, E, and F, you will then find that he has travelled thirty-seven miles in five turnings. Now, how far can he go in fifteen turnings?

The Fly on the Octahedron

"Look here," said the professor to his colleague, "I have been watching that fly on the octahedron, and it confines its walks entirely to the edges. What can be its reason for avoiding the sides?"

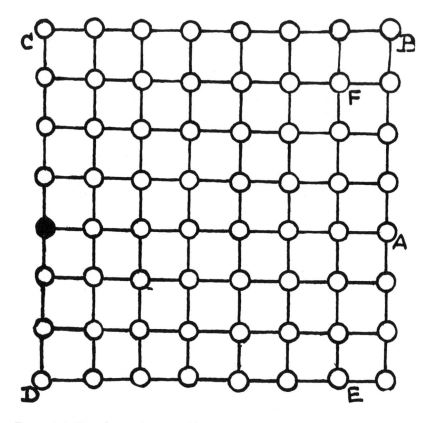

Figure 1.6. How far can he go in fifteen turnings? (Dudeney, p. 70.)

"Perhaps it is trying to solve some route problem," suggested the other. "Supposing it to start from the top point, how many different routes are there by which it may walk over all the edges, without ever going twice along the same edge in any route?"

The problem was a harder one than they expected, and after working at it during leisure moments for several days their results did not agree—in fact, they were both wrong. If the reader is surprised at their failure, let him attempt the little puzzle himself. I will just explain that the octahedron is one of the five regular, or Platonic, bodies, and is contained under eight equal and equilateral triangles. ⌞...⌟ In any route over all the edges it will be found that the fly must end at the point of departure at the top.

"Plenty of Interesting Things to Be Discovered"

NRICH, 1998–2004

The NRICH site at nrich.maths.org, founded in 1996, is one of the web's favorite resources for mathematical activities of every kind, and its collection of games and puzzles grows constantly. Alan Parr's remark about his game, Femto, could well apply to the site as a whole: there are always plenty of interesting things to be discovered.

These problems appear on the NRICH website http://nrich.maths.org and are used with permission.

Sprouts: nrich.maths.org/1208
Noughts and Crosses: nrich.maths.org/1199
Femto: nrich.maths.org/1179 (by Alan Parr)
Nim: nrich.maths.org/402

Sprouts

This is a game for two players. All you need is paper and a pencil. The game starts by drawing three dots. (See Figure 1.7.)

The first player has a turn by joining two of the dots and marking a new dot in the middle of the line. Or the line may start and end on the same dot. (See Figures 1.8 and 1.9.)

When drawing a line, it cannot cross another line. (This is important to remember!) A dot cannot have more than three lines branching to or from it. For example, in Figure 1.10, dots *A* and *B* cannot be used any more because they already have three lines.

The idea is to make it impossible for the other player to draw a line. So the last person to draw a line is the winner. What are the winning tactics? Does it matter who goes first? Why must the game end after a limited number of moves? How many? What happens when you start the game with four or five dots?

Endless Noughts and Crosses

This is a game for two players. You will need a sheet of grid paper (or rule lines down a sheet of writing paper).

Figure 1.7. The beginning of a game of Sprouts.

Figure 1.8. Sprouts: the first move.

Figure 1.9. Sprouts: a different move.

The game is played like the ordinary game of noughts and crosses, with each player taking turns to mark a square with a nought or a cross, but it does not end with first string of three noughts or crosses. Keep going until either the grid is full or both players have had enough! The winner is the player who has the most strings-of-three. You might find it helpful to use different colour pens or to keep score as you play.

Playing on such a large grid means that the game is very unlikely to end in a draw and there is plenty of time to think about strategies for winning.

Once you have mastered strings-of-three, try a game with strings-of-four, then strings-of-five, maybe even strings-of-six!

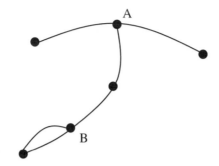

Figure 1.10. Sprouts: dots *A* and *B* are used up.

Femto

There is a book of mathematical games I'm rather fond of, even though it never made the best-seller lists. We called it "What's My Game?", and included in it was a two-player cardgame called Divide and Conquer, for which the entire equipment is a single suit of playing cards. This seemed to me to be just about the most minimal cardgame anyone could devise—until I discovered Pico, which is a German cardgame using just 11 cards.

In case anyone thinks Pico is still just a little too big I've put together ideas from both these games and come up with a new one, called Femto, which needs just eight cards.

Although the game is so small there are plenty of interesting things to be discovered in Femto ⌞...⌟.

Rules for FEMTO

1. Femto is a cardgame for two players; a Femto pack consists of eight cards numbered 2, 3, 4, 5, 6, 7, 8, 10.
2. The cards are shuffled and dealt, so each player gets four cards.
3. In each round of play each player puts out one card, face down. The two cards are then turned face up.
4. The round is won by the higher value card, unless the higher card is more than twice the value of the lower, in which case the lower card wins, e.g., 10 beats 8, 6 beats 5, 3 beats 10, 10 beats 5, ...

5. Whoever played the winning card chooses one of the two cards and puts it, face up, on the table in front of him/her. The player of the losing card takes the remaining card and puts it back into his/her hand.
6. More rounds are played until one player has no cards left.
7. The winner is the player with the greater total value of cards in front of them at the end of the hand.

Well, for a tiny game there seem to be plenty of questions to explore. How long do games last? What is a typical winning score? How many different hands are possible? Which cards are the most powerful? Indeed, what is the most powerful hand (and how could it be beaten)?

Then there's another set of questions, and as a game developer these are the ones which particularly interest me. The first version of any game is often interesting enough, but needs a bit of tinkering to turn it into the finished version. Perhaps the most obvious question to ask is whether we've got the right set of cards. To take another set almost at random, would 5, 7, 9, 13, 16, 18, 20, 24 be any better? Or perhaps we should stick with the original set but add a 12: deal out four cards each so that one card is left unused. Or perhaps there's a role for a Joker card.

Other questions which I'd want to explore concern the game structure. I've suggested that the cards are dealt out, but what if the players take turns to choose cards to make up their starting hand? And instead of playing simultaneously, what if each round is played with one player leading a card face up, so that the second player can see its value before responding? I've also got a hankering to change rule (5), so that it's the losing player who chooses which card goes back into his/her hand. Or perhaps it should be not the winner or loser of the individual round who chooses, but whoever is currently winning (or perhaps losing), who makes the choice, ...

Changing the overall winning criterion often has interesting consequences. The most usual way to do this is to make the winner the person with the lower score rather than the higher. Or in this case, the aim might be to score an odd total, or a total that is a multiple of 3, or a prime number, ...

So it's not true after all that Femto is a game. It's more accurate to say that it may become any one of dozens of potential games. (If there's actually too much choice perhaps we'd be better off looking for a still smaller game, Atto?) I don't know how Femto will end up, but I am excited by the possibility of lots of people trying out different versions ⌴...⌴.

Nim

Challenge

Describe the strategy for winning the game of Nim. The rules are simple. Start with any number of counters in any number of piles. Two players take turns to remove any number of counters from a single pile. The loser is the player who takes the last counter.

Hint

Chance plays no part, and each game must end. The only advantage that either player can possibly have is to start or to play second. To work out how to win you need to start by analysing the "end game," and the losing position to be avoided, and then work back to earlier moves. What should you do if there are only two piles? If there are more piles, what happens if you reduce all the piles to one counter in each?

Solution

This is how you gain control of the game. Make a list of the binary numbers for the number of counters in each pile. Now to have control of the game you want to make the number of 1's in each column even. Whatever your opponent does, when you record the new binary number for the pile that has been changed, there will be an odd number of 1's in one or more of the columns. You will be able to make all the "column sums" even again.

For example, if the numbers of counters in the piles are 7, 6, 4 and 1 you get

$$
\begin{array}{ccc}
1 & 1 & 1 \\
1 & 1 & 0 \\
1 & 0 & 1 \\
1 & & \\
\end{array}
$$

A good move now is to take 4 counters from the pile of 7 which makes the number of 1's in each column even.

~◆~ 2 ~◆~

"Much Necessary for All States of Men": From Arithmetic to Algebra

ARITHMETIC NEVER HAD A EUCLID. TO BE SURE, THERE WERE IMPORTANT systematic accounts of the subject from the earliest times; but none of them acquired anything like the status or the longevity of the *Elements* (for one reason why not, see Charles Hutton's account of the history of arithmetic in Chapter 7). So when the printing press arrived, the field was wide open for individual vernacular writers to dominate for a very long time. It was Robert Recorde (c. 1512–1558) who achieved that status as far as English mathematics writing was concerned, and we begin this chapter with him and some of his many imitators.

From that starting point, we follow a short course in arithmetic and algebra, from addition and subtraction in 1543 through to solving cubic equations nearly 400 years later. Arithmetic itself stays the same, but the stately sixteenth-century dialogue between master and scholar is replaced by a more direct and (sometimes) more user-friendly style of presentation. And, indeed, some of the mathematics that was learned in earlier centuries is now all but forgotten: two of our extracts show the "rule of three," once a major part of basic, everyday mathematics, all but vanished today.

The big change in both style and content, though, was the introduction of algebra into arithmetic teaching. Algebra itself was in its infancy in the sixteenth century, but as its notation developed, its power increased, and its techniques became more widely known, it came to replace verbal rules and case-by-case tricks—like the rule of three—producing a gradual revolution in how numerical techniques were taught and thought about. The end of this story comes in Chapter 9, when we will see something of the philosophy of the "New Math," which introduced abstractions about the nature and properties of arithmetical operations at a very early stage indeed. In this chapter, too, we have the opportunity to reflect on what is gained, and what is lost, by the change from, say Nathan Withy's 1792 "rule of three in verse" to the *Popular Educator*'s 1855 account of completing the square.

Addition and Subtraction

Robert Recorde, 1543

Robert Recorde was one of the first authors of mathematical textbooks in English. Here we present sections from near the beginning of his first work, an arithmetic primer entitled *The ground of artes*, where he explains how to add and subtract, and some of the pitfalls of doing so.

Robert Recorde (c. 1512–1558), *The ground of artes teachyng the worke and practise of Arithmetike, much necessary for all states of men. After a more easyer & exacter sorte, then any lyke hath hytherto ben set forth: with dyuers newe additions, as by the table doth partly appeare.* (London, 1543), fols. 17–19, 37–39.

Addition

⌊MASTER:⌋ Addition is the reduction and bringing of two sums or more into one. As, if I have 160 books in the Latin tongue, and 136 in the Greek tongue, and would know how many they be in all. I must write those ⌊two⌋ numbers, one over another, writing the greatest number highest, so that the first figure of the one be under the first figure of the other, and the second under the second, and so forth in order. When you have so done, draw under them a straight line: then will they stand thus.

$$\begin{array}{r} 160 \\ \underline{136} \end{array}$$

Now begin at the first places, toward the right hand always, and put together the ⌊two⌋ first figures of those two sums, and look what cometh of them; write ⌊it⌋ under them right under the line. As in

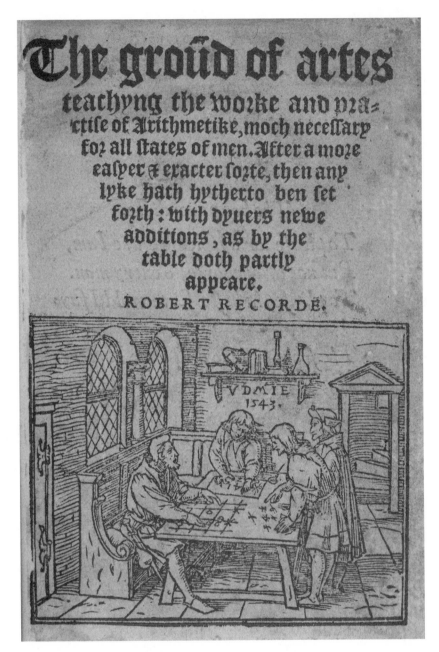

Figure 2.1. The mayster and scolers at work, on the title page of Recorde's *Ground of artes*. ©The British Library Board. G.16099.

saying, "6 and 0 is 6," write 6 under 6, as thus.

$$
\begin{array}{r}
160 \\
136 \\
\hline
6
\end{array}
$$

And then go to the second figures, and do likewise: as in saying, "3 and 6 is 9;" write 9 under 6 and 3, as here you see.

$$
\begin{array}{r}
160 \\
136 \\
\hline
96
\end{array}
$$

And likewise do you with the figures that be in the third place, saying, "1 and 1 be 2:" write 2 under them, and then will your whole sum appear thus.

$$
\begin{array}{r}
160 \\
136 \\
\hline
296
\end{array}
$$

So that now you see that 160 and 136 do make in all 296.

S͵CHOLAR:͵ What, this is very easy to do; me thinketh I can do it, even ͵thus͵. There came through Cheapside ͵two͵ droves of cattle: in the first was 848 sheep, and in the second was 186 other beasts. Those two sums I must write as you taught me, thus.

$$
\begin{array}{r}
848 \\
186 \\
\hline
\end{array}
$$

Then I put the ͵two͵ first figures together, saying, "6 and 8, they make 14." That must I write under 6 and 8, thus.

$$
\begin{array}{r}
848 \\
186 \\
\hline
14
\end{array}
$$

M. Not so, and here are you twice deceived: first in going about to add together ͵two͵ sums of sundry things: which you ought not to do, ͵unless͵ you seek only the number of them, and care not for the things. For the sum that should result of that addition should be a sum neither of sheep nor other beasts, but a confused sum of both. How be it, sometimes you shall have sums of diverse denominations to be added, of which I will tell you anon. But first I will show you where you were deceived in another point, and that was in writing 14, which came of 6 and 8, under 6 and 8, which is impossible: for how can two figures of two places be written under one figure and one place?

S. Truth it is, but yet I did so understand you.

M. I said indeed that you should write that under them, that did result of them both together, which saying is always true, if that sum do not exceed a digit. But if it be a mixed number, then must you write the digit of it under your figures, as I have said before. But and if it be ‚ten‚, then write 0 under them, and keep the ‚ten‚ in your mind. And therefore when you have added your second figures, which occupy the place of tens, you shall put that 1 thereto, which you kept in your mind.

Subtraction

S. Then have I learned the two first kinds of Arithmetic; now (as I remember) doth follow Subtraction, whose name me thinketh doth sound contrary to Addition.

M. So is it indeed: for as Addition increaseth one gross sum, by bringing many into one, so contrariwise Subtraction diminisheth a gross sum by withdrawing of ‚an‚ other from it; so that Subtraction or rebating is nothing else, but an art to withdraw and abate one sum from another, that the remainder may appear.

S. What call you the remainder?

M. You may perceive by the name.

S. So me thinketh, but yet it is good to ask the truth of all such things, lest in trusting to my own conjecture I be deceived.

M. So is it the surest way. And as I see cause, I will still declare things unto you so plainly that you shall not need to doubt. Howbeit, if I do overpass it sometimes (as the manner of men is to forget the small knowledge of them to whom they speak) then do you put me in remembrance yourself, and that way is surest. And as for this word that you last asked me, take you this description: The remainder is a sum left, after due working, which declareth the excess or difference of the two other numbers. As, if I would deduct 14 out of 18, there should remain 4, which is called the remainder, and is the difference between those two numbers 14 and 18.

S. I perceive then what subtraction is. Now resteth to know the art to work by it.

M. That shall you do by this means: first you must consider that if you should go about to rebate, you must have two sundry sums proposed. The first, which is your gross sum or sum total, and it must be set highest, and then the rebatement, or sum to be withdrawn, which must be set under the first (whether it be in one parcel or in many), and that so that the first figures be one just over another, and so the second, and third, and all other following, as you did in Addition. Then shall you draw under them a line: and so are your sums duly set to begin your working.

Then begin you at the right hand (as you did in Addition) and withdraw the nether number out of the higher. And if there remain anything, write that right under them beneath the line, and if there remain nothing (by reason that the two figures were equal) then write under them a cipher of nought. And so do you with all the other figures, ever more abating the nether out of the higher, and write under them the remainder still, till you come to the end.

And so will there appear under the line what remaineth of your gross sum, after you have deducted the other sum from it, as in this example. I received of your father 48 shillings, of which I have laid out for you 36 shillings. Now would I know what doth remain, and therefore I set my numbers thus in order. First I write the greatest sum, and under him the lesser, so that the figures at the right side be even, one under another, and so the others thus.

$$48$$
$$\underline{36}$$

Then do I rebate 6 out of 8, and there resteth 2, which I write under them right beneath the line, thus.

$$48$$
$$\underline{36}$$
$$2$$

Then I go to the second figures, and do rebate 3 out of 4, where there remaineth 1, which I write under them right, and then the whole sum and operation appeareth thus.

$$\begin{array}{r} 48 \\ 36 \\ \hline 12 \end{array}$$

Whereby it appeareth that if I withdraw 36 out of 48, there remaineth 12.

Multiplication and Division

Thomas Masterson, 1592

The mathematical books by Robert Recorde and Humfrey Baker (whose "Sports and pastimes, done by number" we met in Chapter 1) were reprinted many times, and they dominated the English market for arithmetic primers during the second half of the sixteenth century and well into the seventeenth. Competition tended to be indirect, and in Thomas Masterson's *First book of Arithmetic* (there were eventually three), we see a volume aimed at rather more gentlemanly readers, complete with a dedication to a nobleman and a quasi-Euclidean set of definitions and "declarations" at the beginning of the book. Despite this, the title page echoes Recorde's claim—or rather advertising slogan—that arithmetic was "very necessary" for "all men."

Masterson's prose is paraphrased here, and his examples expanded.

Thomas Masterson (fl. 1592–1595), *Thomas Masterson his First Booke of Arithmeticke. Shewing the ingenious inuentions, and figuratiue operations, by which to calculate the true solution or answeres of Arithmeticall questions: after a more perfect, plaine, briefe, well ordered Arithmeticall way, then any other heretofore published: verie necessarie for all men.* (London, 1592), pp. 7–10.

Of Multiplication

Two numbers being given, of any kinds whatsoever, to find a third number, which shall contain the one as many times as the other contains a unit.

Place the numbers as if they were to be subtracted. Multiply the first figure (starting from the right) of the lower number by the first of the upper. If their product is no more than nine, write it just under them. But if it is more than nine, write only the first figure of it, and add the other figure to the number which comes from the first figure of the lower number, multiplied by the second of the upper. And with that sum, do as you did with the last number—except that the figure must be written next to the figure last written—adding the second figure, if it amounts to more than nine, to that which comes from the first figure of the lower number multiplied by the next figure of the upper.

In the same way, continue, until all the figures of the upper number are multiplied by the first of the lower. Then do with the second, third, and all the other figures of the lower number, in order, as you have done with the first. But when you begin to multiply with a new figure of the lower number, begin to write the result under that figure.

And then the sum of all those new-found numbers, being added together, is called the product, and is the number which comes from the two numbers given, being multiplied together. It contains one of them as many times as the other contains a unit.

Examples

$$
\begin{array}{r}
784378 \\
\times\ 987 \\
\hline
5490646 \\
6275024 \\
7059402 \\
\hline
\end{array}
$$

makes   774181086

$$
\begin{array}{r}
984723 \\
\times\ 3 \\
\hline
\end{array}
$$

makes   2954169

$$
\begin{array}{r}
890084 \\
\times\ 5 \\
\hline
\end{array}
$$

makes   4450420

Of Division

Two numbers of one kind being given, to find how many times the lesser number is contained in the greater; and also how much will remain, if the lesser is subtracted from the greater as many times as possible.

Place the figure which stands in the first place of the lesser number (that is, of the divider) under the figure which stands in the first place of the greater number (that is, of the dividend), and then all the rest in order following. But if the first figure of the divider is greater than the first figure of the dividend, or if the divider is larger than the number made by the figures of the dividend which stand over it, taken on their own, then write the first figure of the divider under the first but one of the dividend, and the others in order after it.

Then write, separately, how many times the first figure of the divider is contained in the figure (or, if necessary) figures that are over it, taken separately. Then multiply the divider by this figure which you have written separately apart, and subtract the result from the dividend.

Then place the divider one figure further left, and repeat the procedure. When you write, separately, how many times the first figure of the divider is contained in the figure(s) over it, you must append it to the right of the previous such figure.

Continue in the same manner, moving the divider, until the figures in the last places of the divider and dividend stand under one another. Then the figures which are written separately are called the quotient, and they show how many times the divider is contained in the dividend. What remains after the last subtraction shows how much remains when the divider has been taken out of the dividend as many times as it can be.

Example

978456 to be divided by 8.

First digit: 9. The divisor, 8, goes into 9 once, so the first digit of the answer is 1. And $9 - 8 = 1$, so the new dividend is 178456.

First digits: 17. The divisor goes into 17 twice, so the second digit of the answer is 2. And $17 - 16 = 1$, so the new dividend is 18456.

First digits: 18. The divisor goes into 18 twice, so the third digit of the answer is 2. And $18 - 16 = 2$, so the new dividend is 2456.

First digits: 24. The divisor goes into 24 three times, so the fourth digit of the answer is 3. And $24 - 24 = 0$, so the new dividend is 56. (Since two digits of the dividend have been removed, we shift the divisor two places to the right, and we put a zero as the fifth digit of the answer.)

First digits: 56. The divisor goes into 56 seven times, so the sixth digit of the answer is 7. And $56 - 56 = 0$, so there is no remainder.

The answer, in full, is 122307.

Reducing Fractions

John Tapp, 1621

John Tapp's *Path-Way to Knowledge* is a clear imitation of Robert Recorde's works in both its title and its dialogue style of presentation, with "Theodore" and "Junius" replacing Recorde's "Master" and hapless "Scholar." Tapp was best known as the compiler of a nautical almanac, *The Seaman's Kalendar*, and as the author and publisher of other works on navigation, a subject which he also taught privately. Where Recorde's *Whetstone of Witte* had been dedicated to the Muscovy Company, Tapp dedicated this book to the same company's governor, Sir Thomas Smyth, reflecting a continuing—and justified—belief in the practical importance of arithmetical skills in the world of international trade. Here we see an explanation of what fractions are and how to "take them up": that is, reduce them to their lowest terms.

John Tapp (c. 1575–1631), *The Path-Way to Knowledge; Contayning the whole Art of Arithmeticke, both in whole Numbers, and Fractions; with the extraction of Roots; as also a briefe Introduction or entrance into the art of Cossicke Numbers, with many pleasant questions wrought thereby ⌊ . . . ⌋* (London, 1621), pp. 85–89.

How a Fraction is expressed

THEODORE. A Fraction, indeed, is a broken number, and so consequently the part of another number. But that must be understood, of such another number as cannot be divided into any other parts than fractions. Example: although I may take several parts of this number, 24—as the one-half is 12, the one-third part is 8, the fourth part is 6, and so many other parts diversely—yet these parts are not, nor ought not properly, to be called fractions, because they may be expressed by whole numbers. But a fraction properly expresseth the parts or part of a unity. That is to say, the entire sum that a fraction doth express, it cannot be so great that it shall make 1. Therefore you shall understand the expressing of a fraction is represented by two numbers set one over the other, and a line drawn between them, thus: $\frac{1}{2}, \frac{2}{3}, \frac{3}{4}, \frac{5}{6}$, which 4 fractions are to be pronounced thus: one half, two third parts, three fourth parts, five sixth parts.

JUNIUS. I understand how they are expressed and their pronunciation, but of their values I am uncertain. But I think they are thus valued: if a pound of money be divided into 2 parts, that first fraction, $\frac{1}{2}$, doth express one of those 2 parts. And the latter fraction, $\frac{5}{6}$, doth signify, if a pound be divided into 6 parts, I must know it, to be 5 of those 6 parts. And so consequently I conceive of the rest or any such like. If there be no more difficulty in expressing or numbering of a fraction, I pray you proceed forward.

THEOD. There is no more difficulty, but only that you express the names aright of both numbers which maketh a fraction. The overmost, which is above the line, is called the Numerator, and the other, beneath the line, is called the Denominator. The reason is, the overmost doth express the Numerator or number of parts that the fraction doth contain; the denominator and nether number expresseth the Denomination or name of parts whereinto the unity or whole thing is divided.

JUNI. Are there no other kinds of Fractions which you have not yet taught?

THEOD. There are of Fractions (or that are expressed as fractions) 4 kinds, whereof 2 kinds of them are properly fractions, and the other 2 kinds are not properly fractions, but are commonly so expressed. This first kind which I have now showed you is truly a fraction, and before I meddle with any of the rest, I will show you how to take up any fraction that shall remain in any division, when you work in whole numbers.

Abbreviation of Fractions

THEOD. ... You must consider what part you may take, or how you may divide both Numerator & Denominator. If both numbers are even, as this fraction is, $\frac{294}{336}$, then you must practise this division by even parts, either by $\frac{1}{2}, \frac{1}{4}, \frac{1}{6}, \frac{1}{8}, \frac{1}{12}$, or more even parts, until such time as you cannot take an even part out of both of them alike, but that there will remain some odd number. Then seek some odd part or number that will divide them both, as $\frac{1}{5}, \frac{1}{7}, \frac{1}{9}, \frac{1}{11}$, or more odd parts. But if you can find no such part to be taken of them, whereby they cannot be brought to a lower denomination, then you must name them as you have found them.

Example: $\frac{294}{336}$. Take $\frac{1}{2}$ of 294, the Numerator, which is 147, and $\frac{1}{2}$ of 336, being the Denominator, which is 168. Then for $\frac{294}{336}$ I find $\frac{147}{168}$, and both of one value.

Because the numerator 147 is an odd number, I seek to take some odd part both of the Numerator and Denominator, where I do consider the $\frac{1}{7}$ part of either of them may be taken. The $\frac{1}{7}$ of 147 the Numerator is 21; the $\frac{1}{7}$ of 168 the denominator is 24: then for $\frac{147}{168}$ I find $\frac{21}{24}$, and both of one value.

It is an odd number, ⌊and⌋ you must yet divide the fraction. I do consider $\frac{1}{3}$ part will divide them both. The $\frac{1}{3}$ part of the Numerator, 21, is 7, and the $\frac{1}{3}$ part of the Denominator, 24, is 8; then for $\frac{21}{24}$ I find $\frac{7}{8}$, so that $\frac{294}{336}$ of a pound is become $\frac{7}{8}$, which is in value 17 shillings, 6 pence. And there it must rest, because it cannot be expressed in any lower term.

JUNI. I have the understanding of this manner of the taking up of a fraction, as you shall understand by an example or two in fractions, which I will express and work myself after the manner you have showed me examples, if you would have me abbreviate these fractions.

$$\frac{544}{612} = \frac{272}{306} = \frac{136}{153} = \frac{8}{9}$$

$$\frac{693}{847} = \frac{63}{77} = \frac{9}{11}$$

I doubt not but that I have here wrought truly, and I am sure I can take up any fraction after this manner; but now they are thus taken up, I know not their certain values no more than I did before.

THEOD. You have well done in Abbreviating your fraction, if you understand why they are thus abbreviated.

JUNI. I am showed before what it is to express a fraction in his lowest term⌊s⌋, or to reduce him into his lowest denomination. And I do also conceive that his value is neither augmented nor diminished, but that $\frac{8}{9}$, the lowest term, is equal to $\frac{544}{612}$, the highest term, and $\frac{9}{11}$ is equal to $\frac{691}{847}$: and so consequently I am to conceive and judge of all fractions that shall be so taken up.

Decimal Fractions

Edward Hatton, 1695

Like Masterson and Tapp in the previous two extracts, Edward Hatton was yet another individual whose mathematical writing was conditioned by the dominating presence of Robert Recorde's works, even a century and a half after their first appearance. He published a new edition of Recorde's *Ground of Artes* and kept a mathematical school in Worcestershire: as the title page of this book shows, his interests were particularly in the mathematics of trade.

Decimal fractions had been introduced early in the seventeenth century by the Dutch engineer Simon Stevin, and by this date the primer on, specifically, decimal arithmetic had established itself as a separate genre, side by side with its "vulgar" cousin.

Edward Hatton (fl. 1696–1714), *The Merchant's Magazine: or, Trades-man's Treasury. Containing Vulgar Arithmetick in Whole Numbers, with the Reason and Demonstration of each Rule, adorn'd with curious Copper Cutts of the chief Tables and Titles: Also Vulgar and decimal Fractions, after a New, Easie and Practical Method. Merchants Accompts, or Rules of Practice ⌊...⌋. Book-keeping, after a Plain, Easie and Natural Method ⌊...⌋. And Lastly, Maxims to be observed in Drawing, and Accepting Bills of Exchange ⌊...⌋.* (London, 1695), pp. 76–78.

A Decimal Fraction is only different from a Vulgar in this: That the Denominator of a Decimal Fraction is either 10, or some power of 10, *viz.* 100, 1000, 10000, *etc.*, so that the Denominator is easily known without expressing it. For in a Decimal Fraction there is a Point or Prick toward the Left-hand of the Numerator, which Point always possesses the like place, as the first Figure toward the Left-hand would, if it were to be wrote down. Thus $\frac{1}{10}$ is .1, the Prick being in the Tens place, and therefore denotes the Denominator to be 10; $\frac{12}{100}$ is .12; $\frac{125}{1000}$ is .125; $\frac{1964}{10000}$ is .1964; $\frac{17}{1000}$ is .017; $\frac{24}{10000}$ is .0024, etc. The manner to reduce a Vulgar Fraction to a Decimal, is by this Proportion.

Rule

As the Denominator of the Vulgar Fraction given is in proportion to its Numerator, so is 1000 to the Numerator of the Decimal, whose

Denominator is 1000. Or: ... so is 10,000 to the Decimal whose Denominator is 10,000. Etc.

Example

What is $\frac{1}{8}$ in a Decimal Fraction? See the Operation.

⌞8 : 1 = 1000 : ?
1000 × 1 ÷ 8 = 125
Answer: .125⌟

But because it sometimes happens that a Cipher or more is to possess the 1, 2, etc., Places of the Decimal toward the Left-hand, therefore take this General Rule:

As many Ciphers as you have in the third Number of the 3 in ⌞the⌟ proportion as above, so many Places must you prick off in the Quotient toward the Right-hand.

Example 2

How is 9 pence expressed in the Decimal of a Pound Sterling?

Rule

Consider that in a Pound are 240 Pence, therefore 9 pence is $\frac{9}{240}$ pounds in a vulgar Fraction. ⌞...⌟ Then say, as in the last Example:

⌞240 : 9 = 10000 : ?
10000 × 9 ÷ 240 = 3750
Answer: .0375 pounds.⌟

In this Example, because I had 4 Ciphers in the third Number, therefore I must prick 4 places off toward the Right-hand the Quotient for Decimals. But because the said Quotient did but consist of 3 places, therefore I supply the fourth to the Left-hand with a Cipher.

Note that the greater your third Number is, the nearer do you bring your Decimal to Truth, when anything happens to remain ⌞...⌟. But in most Cases where the Decimal is not to be multiplied by a great Number, it is sufficient that the fourth Number be 1000.

But when you reduce $\frac{1}{4}$ or $\frac{1}{2}$ or $\frac{3}{4}$ to Decimals, or any Number of shillings to the Decimal of a Pound, it is sufficient in these Cases if your third Number be 100.

Extracting Square Roots

William Banson, 1760

Manuals of arithmetic were numerous in the eighteenth century, and William Banson's is a typical example. He was a minor figure, who worked as a teacher of handwriting and an accountant; this book was his one foray into mathematical exposition. His other published works were a set of currency conversion tables and a volume of specimens of good handwriting.

His book was tailored to practical use by teachers, and some of the "examples" were in fact blank spaces, to be filled in with whatever examples the teacher pleased.

William Banson (fl.?1717–1760), *The School-Master and Scholar's Mutual Assistant: Or, a Compendious System of Practical Arithmetic, Made perfectly Easy*. (London, 1760), pp. 145–146.

Extraction of the Square Root

When any Number is multiplied by itself, the Product is called the Square of that Number, and the Number itself is called the Square Root of that Product. So that ₍since₎ 3 multiplied by 3 gives 9, therefore 9 is the Square of 3, and 3 is the Square Root of 9; also 36 is the Square of 6, and 6 is the Square Root of 36; etc. In like Manner every Number squared, and multiplied again by itself, produces the Cube of that Number. As may be seen in the following Table, wherein the Squares and Cubes are placed together.

A Table of Squares and Cubes, and their Roots.

Root.	1	2	3	4	5	6	7	8	9
Squ₍are₎.	1	4	9	16	25	36	49	64	81
Cub₍e₎.	1	8	27	64	125	216	343	512	729

To extract the Square Root of any Number, you must

1st, Begin at the Units Place, and make a Point or Period over it, and also over every second Figure.

2ndly, Take the nearest less Square Root of the first Period, (i.e. the first one or two figures) towards the Left-hand, (which may be easily found by the foregoing Table) and place it like a Quotient Figure in common Division.

3rdly, Subtract its Square from the said first Period, and to the Remainder bring down the next Period, or 2 Figures, and call that the Resolvend.

4thly, Double the Root or Quotient Figure, and place it for a Divisor on the Left-hand of the Resolvend, and seek how often the Divisor is contained in the Resolvend, ignoring the Units Place; which Figure place in the Quotient, and likewise in the Units Place of the Divisor. Then multiply this Divisor by the last Figure put in the Quotient, and subtract the Product from the Resolvend (as in common Division).

And so proceed to work in like Manner for each Period that is in the given Number.

Note: There must be just so many Figures in the Quotient, and so many Operations, as there are Points in the given Number.

Example I

Let 2304 be a Number given, and let the Square Root thereof be required.

$$2\overset{\centerdot}{3}0\overset{\centerdot}{4} \text{ (48 Root required.}$$

$$\begin{array}{r} 16 \\ \hline \end{array}$$

Divisor 8) 704   Resolvend.

Divisor 88

$$\begin{array}{r} 704 \\ \hline \end{array} \quad \text{Product.}$$

Instruction

In the Example above, I first point the given Number as before directed, putting a Point upon the Units, and another upon the Hundreds.

Then I seek the greatest square Number in 23, which I find to be 16, 4 being the Root thereof. So that I place 16 under 23, the first Period, and 4 in the Quotient. Then I subtract 16 from 23, and there remains 7, to which Remainder I bring down the next Period, and place it on the Right-hand, which makes 704 for a Resolvend.

Then I double the Quotient, 4, and it makes 8, which I carry to the Left-hand of the Resolvend for a Divisor, and seek how many Times 8 will go in 70. The Answer is 8, which I put in the Quotient, and also place it on the Right-hand of the Divisor, which makes it 88.

Then I multiply 88 by the 8 I put in the Quotient, and the Product is 704; which subtract from the Resolvend, and nothing remains. So that the Work is done, the Square Root of 2304 being 48.

The Rule of Three

Wardhaugh Thompson, 1771

I couldn't resist including this manual of accounting, whose (otherwise unknown) author is the only individual I have ever come across to have my surname as his first name. The extract shows the rule of three, a regular feature of arithmetic books until at least the nineteenth century. It amounted to the rule that if $a : b = c : d$ then $d = cb/a$, so that if a proportionate relationship holds between four quantities we can always, knowing three of them, find the fourth (if a caterpillars will eat b leaves in a day, we can find out how many leaves c caterpillars will eat in a day: d). This, and more complex variants with names like the "double rule of three" and the "rule of three inverse" had many practical applications, and they occupied, perhaps, an intermediate position between the numerical specificity of arithmetic and the abstraction and generalization of algebra.

Wardhaugh Thompson, many years an accomptant in London, *The Accomptant's Oracle; or Key to Science. Being a Treatise of Common Arithmetic: With the Doctrine of Vulgar and Decimal Fractions. Upon a Plan Entirely New. To which are added Decimal Tables, with their Use and Construction.* (Whitehaven, 1771), pp. 84–86.

Of the Rule of Three

Observation, 1st. The Rule of Three is either single or compound.

2nd. The single Rule is: when three terms are given, to find a fourth.

3rd. Of the three terms given, two of them always imply a supposition, and the third is a demand. So, this question being proposed, *viz.*,

If 1 pound of sugar cost 4 pence, what will 6 pounds cost at that rate?

Here it is plain that the 1 pound of sugar and the 4 pence are the two terms of the supposition, and the 6 pounds of sugar is that which demands or asks the question. ⌐The question⌐ is generally known by following these (or suchlike) words: *How many? How much? What will? How long? How far?* etc.

4th. One of the terms in the supposition is always of the same name with that which asks the question, and the other term of the supposition ⌐is⌐ of the same kind with the fourth term, ⌐which is the⌐ answer required.

5th. The Rule of Three is also either *Direct*, or *Inverse*.

6th. The Rule of Three direct is when more requires more, or less requires less. That is, when (according to the sense and tenor of the question) the third term is more or greater than the first, and requires the fourth term ⌐to be⌐ more or greater than the second in the same proportion. Or when the third term is less than the first, and requires the fourth term or answer ⌐to be⌐ less than the second in the same proportion. For as often as the first term contains the second, or is contained by the second, just so often must the third contain the fourth, or be contained by the fourth.

7th. ⌐Given⌐ four numbers which are proportionals, the product of the two extremes (which are first and fourth) will be equal to the product of the two means, which are the second and third.

8th. In stating all questions in this rule, let that term of your supposition which is of the same name with that term of the demand, be your first. ⌐Let⌐ the other term in your supposition, which will be of the same name with the answer required, ⌐be⌐ your second. And consequently ⌐let⌐ that of your demand ⌐be⌐ the last. And then the question will stand in the following order.

	⌐pounds	pence		pounds	pence⌐
If	1	: 4	⌐=⌐	6	: ?⌐

To be read thus. If 1 pound cost 4 pence, what will 6 pound cost?

And here, according to the sixth observation aforegoing, I find that more requires more: for if the price of one pound be 4 pence, the price of 6 pound will consequently be 6 times as much.

I shall next give you an example where less requires less, which may be the following, *viz.*

If 6 pounds of sugar cost 24 pence, what will 1 pound cost? Stated thus:

	pounds		pence		pounds		pence
If	6	:	24	=	1	:	?

In this question (according to the sixth observation aforesaid) less requires less: for if 6 pounds cost 24 pence, one pound must therefore cost six times less, *viz.*, 4 pence.

After having stated your question as afore directed, to perform the operation, this is the Rule:

Multiply your second and third terms together, and divide their product by the first; and the Quotient will be the answer.

The Rule of Three, in Verse

Nathan Withy, 1792

Nathan Withy's *A Little Young Man's Companion* is one of the more attractive of the mathematics primers produced during the eighteenth century. Withy was the author of several short books in verse, ranging from the satirical to the moralistic, but his writing career never really seems to have taken off, and this was his only foray into mathematics.

Here we see how he deals with the rule of three (compare Wardhaugh Thompson's version in the previous extract).

Nathan Withy (fl. 1777–1795), *A Little Young Man's companion; Or, Common Arithmetic, Turned into a Song, As far as the Rule of Three Direct. Written for the Benefit and Instruction of those who have not Time to read large Books. To which is added, One Enigma, A New Song in Praise of London Porter, and The Wandring Bard's Farewel to Oxford. By N. Withey, of Hagley, in Worcestershire.* (London, 1792), pp. 9–10.

The Golden rule has always been
    Composed of numbers three.
These stated right will find a fourth,
    Shall in proportion be;

The fourth and second brothers are,
 Believe me on my word,
Either men, money, *et cetera*;
 So are the first and third.
Multiply the second by the third,
 And write their product fair,
And then divide it by the first,
 With diligence and care;
Their quotient is the answer then,
 And as such it will agree,
For 'tis the number called the fourth,
 Produced by t'other three.
To prove it, I will tell how,
 With pencil, pen, or feather,
For you must multiply the first
 And fourth numbers together,
And if their product is the same
 With the second and the third,
You may conclude your work is right,
 Believe me on my word.
This by example I will prove:
 Suppose that two is three,
I beg you'll tell by the same rule,
 What five will come to be;
You'll find it seven and a half,
 As plainly may be seen,
And this answer multiplied by two,
 Will turn out just fifteen.
This is the rule that gunners use,
 With diligence and care,
To throw their bombs to any spot,
 Or mount them in the air;
By it ten thousand things are done,
 Ten thousand different ways,
And he that learns it perfectly,
 Will merit fame and praise.

"The First Analysts"

Joseph Fenn, 1775

Joseph Fenn worked at the University of Nantes, ran a school in Dublin, and also published a version of Euclid. Here we see him speculating about the historical origins of algebra (his speculations apparently untainted by any acquaintance with historical facts). Algebra using symbols had existed since the sixteenth century, and much modern notation since the seventeenth, but it was slow to become part of a basic mathematical education; for many readers specific rules like the rule of three of the previous two extracts rendered abstract algebraic manipulation unnecessary. Fenn shows us how such practical rules could have evolved into the solving of equations.

Joseph Fenn (fl. 1769–1783), *A New and Complete System of Algebra: or, Specious Arithmetic. Comprehending All the Fundamental Rules and Operations of that Science, clearly Explained and Demonstrated, with the Resolution of all kinds of Equations, Illustrated and Exemplified in the solution of a Vast Variety of the Most Curious and Interesting Questions. For the Use of Schools.* (Dublin, 1775?), pp. 1–2, 119–120.

Of the analytic Method of expressing Problems by Equations, and of the Resolution of Equations of the first degree

Amongst the different Problems which employed the first Mathematicians called Analysts, I choose the following, as the most proper to show how they formed the Science styled specious Arithmetic.

I. To divide a Sum, for Example, £890, between three Persons, in such a Manner that the first may have £180 more than the second, and the second £115 more than the third

It is thus I imagine a Person would have argued, who, without the least Tincture of specious Arithmetic, attempted to solve this Problem.

　　It is manifest that if one of the three Parts was known, the other two would be immediately discovered. Let us suppose, for Example, the third, which is the least, to be known; we must add £115 to it, and this Sum will be the Value of the second; to obtain afterwards the first Part we must add £180 to this second, which comes to the same as if we added £180 ₍plus₎ £115, or £295, to the third.

Let therefore this third Part be what it will, we know that this Part, plus, itself together with £115, plus, itself again together with £295, should make a sum equal to £890.

From whence it follows that the Triple of the least Part, plus, £115 plus, £295, or plus, £410, is equal to £890.

But if the triple of the Part sought, plus, £410, be equal to £890, this Triple of the Part sought must be less than £890 by £410. Therefore it is equal to £480, therefore the least part is equal to £160. The second will consequently be £275, and the first or greatest £450.

It is probable the first Analysts argued in this Manner when they proposed to themselves questions of this Nature. Without doubt, in proportion as they advanced in the Solution of a Problem, they burdened their Memories with all the Arguments which had conducted them to the Point they had arrived at, and when the Problems were not more complicated than the foregoing, it was no difficult Matter. But as soon as their Researches presented a greater Number of Ideas to be retained, they were under the Necessity of having recourse to a more concise Method of expressing themselves, and of employing some simple Symbols, by Means of which, however advanced they were in the Solution of a Problem, they might perceive at one View what they had done and what remained for them to do. Now the Kind of Language they imagined for this Purpose, is called specious Arithmetic.

II

To explain the Principles of this Science more clearly, we will resume the same Question, write down in Words the Arguments which the Analyst employs to solve his Problem, and in analytic Symbols what is requisite to assist his Memory.

The least or third Part, be it what it will, I denote by one Letter, for Example by x.

The second consequently will be x plus, 115, which I denote thus: $x + 115$, employing the Sign + which signifies *plus*, to express the Addition of the two Quantities between which it is placed.

As to the first Part or greatest, since it exceeds the second by 180, it will be expressed by $x + 115 + 180$.

Adding those three Parts we will have $3x + 115 + 115 + 180$, or, when reduced, $3x + 410$.

But this sum of the three Parts should be equal to 890, which I express thus: $3x + 410 = 890$, employing the Symbol =, which signifies *equal*, to denote the Equality of the two Quantities between which it is placed.

The Question therefore, by this Computation, is changed into another, where it is required to find a Quantity, the Triple of which being added to 410 makes 890. To find the Resolution of similar Questions is what is understood by solving an Equation. The Equation in the present Case is $3x + 410 = 890$; it is so called because it indicates the Equality of two Quantities. To solve this Equation is to find the Value of the unknown Quantity x from this Condition: that its Triple ,plus, 410 makes 490.

III

To solve this Equation the Analyst argues and writes down his Arguments as follows. The Equation to be solved, $3x + 410 = 890$, teaches us that we are to add 410 to $3x$ to make up the Sum 890. Wherefore $3x$ are less than 890 by 410, which I express thus: $3x = 890 - 410$, employing the sign $-$, which signifies *less*, to denote that the Quantity which it precedes should be subtracted from that which it follows.

From this new Equation, $3x = 890 - 410$, we deduce, by subtracting in effect 410 from 890, this other Equation: $3x = 480$.

But if three x be equal to 480, one x will be the third Part of 480, or 160, which I write down thus: $x = \frac{480}{3} = 160$. And the Question is solved, since it suffices to know one of the Parts to discover the rest.

Quadratic Equations

The Popular Educator, 1855

The editors of the six-volume *Popular Educator*—a general encyclopedia of the 1850s, later reissued and revamped several times—boasted in this final volume that it provided "a vast amount of solid and useful information in a popular form, and at a price unprecedented even in the present age of Cheap Literature," claiming that it had benefited "a host of students." One of its aims seems to have been to enable students to feign a classical education who had had none. It had predecessors in the popular encyclopedias and educational serials of the eighteenth century, some of which feature elsewhere in this anthology; but the choice of subjects had, perhaps, a distinctively Victorian flavor: arithmetic, algebra, trigonometry, geology, and physics rubbed shoulders with languages ancient and modern, with reading and elocution, and with "moral science."

The Popular Educator, vol. 6 (London, 1855), pp. 514–515.

Equations are divided into classes, which are distinguished from each other by the power of the letter that expresses the unknown quantity. Those which contain only the *first* power of the unknown quantity are called *simple* equations, or equations of the first degree. Those in which the highest power of the unknown quantity is a *square*, are called *quadratic*, or equations of the *second degree*; those in which the highest power is a *cube* are called *cubic*, or equations of the *third degree*; etc.

Thus $x = a + b$ is an equation of the *first* degree.

$x^2 = c$, and $x^2 + ax = d$ are *quadratic* equations, or equations of the *second* degree.

$x^3 = h$, and $x^3 + ax^2 + bx = d$ are *cubic* equations, or equations of the *third* degree.

❧

In the equation $x^2 + 2ax = b$, the side containing the unknown quantity is not a complete square. The two terms of which it is composed are indeed such as might belong to the square of a binomial quantity.° But one term is *wanting*. We have then to inquire in what way this may be supplied. From having *two* terms of the square of a binomial given, how shall we find the *third*?

Of the three terms, two are complete powers, and the other is twice the product of the roots of these powers, or, which is the same thing, the product of one of the roots into twice the other.

In the expression $x^2 + 2ax$, the term $2ax$ consists of the factors $2a$ and x. The latter is the unknown quantity. The other factor $2a$ may be considered the *coefficient* of the unknown quantity, a coefficient being another name for a factor. As x is the root of the first term x^2, the other factor $2a$ is *twice* the root of the third term, which is wanted to complete the square. Therefore *half* of $2a$ is the root of the deficient term, and a^2 is the term itself.

The square completed is $x^2 + 2ax + a^2$, where it will be seen that the last term a^2 is the square of half of $2a$, and $2a$ is the coefficient of x, the root of the first term.

In the same manner it may be proved that the last term of the square of any binomial quantity is equal to the square of half the coefficient of the root of the first term.

From this principle is derived the following

Method for completing the square
Take the square of half the coefficient of the first power of the unknown
  quantity, and add it to both sides of the equation.

After the square is completed, ⌊quadratic equations⌋ are reduced in the same manner as pure equations.

⌊*Example*⌋ *1. Reduce the equation $x^2 + 6ax = b$*

Completing the square, $x^2 + 6ax + 9a^2 = 9a^2 + b$.
Extracting the root of both sides, $x + 3a = \pm\sqrt{9a^2 + b}$.
And $x = -3a \pm \sqrt{9a^2 + b}$.

Here the coefficient of x, in the given equation, is $6a$. The square of half this is $9a^2$, which being added to both sides completes the square. The equation is then reduced by extracting the root of each member.

Note

binomial quantity: an algebraic expression containing two terms, like $1 + x$ or $pq + rs$.

Cubic Equations for the Practical Man

J. E. Thompson, 1931

The solution of the cubic equation had long been known, but its practical uses remained somewhat elusive, at least at an elementary level. James Edgar Thompson, author of various textbooks including a historically based account of the slide rule, evidently believed that the "practical man" had at least some interest in mathematical results for their own sake.

J. E. Thompson (b. 1892), *Algebra for the Practical Man (Mathematics for Self-Study)*. (London, 1931), pp. 153–158.

The Complete Cubic Equation

The complete cubic equation with one unknown quantity is one which contains a constant term and the first, second and third powers of the unknown but no higher power. When all terms are transposed to the left member and like terms combined, the equation may then be divided by the coefficient of the cube of the unknown and so reduced to the form

$$x^3 + ax^2 + bx + c = 0. \tag{2.1}$$

This will be referred to as the standard form of the *complete cubic equation* in one unknown.

If in this equation a new variable u is substituted for x by putting $x = u - \frac{a}{3}$, where a is the coefficient of the x^2 term, and the resulting equation then simplified, a new equation results, which is of the form

$$u^3 + pu + q = 0. \tag{2.2}$$

In this equation p and q are new coefficients which are made up of combinations of the original coefficients, a, b and c. It is to be noted that there is no second power term in this equation. The second power is said to have been *suppressed* and the equation 2.2 is called the *reduced cubic equation*.

As an example of this reduction we will suppress the square term in the complete cubic equation

$$x^3 - 9x^2 + 9x - 8 = 0, \tag{2.3}$$

in which $a = -9$, $b = 9$, $c = -8$. To do this we put $x = u - a/3 = u - (-9/3)$, or

$$x = u + 3.$$

The equation then becomes

$$(u + 3)^3 - 9(u + 3)^2 + 9(u + 3) - 8 = 0;$$

cubing and squaring the binomial $u + 3$, this is

$$u^3 + 9u^2 + 27u - 9u^2 - 54u - 81 + 9u + 27 - 8 = 0,$$

and when like terms are combined there results finally,

$$u^3 - 18u - 35 = 0, \qquad (2.4)$$

in which there is no u^2 term.

This new cubic equation with u as the unknown quantity is more easily solved than the original equation 2.3. When this equation has been solved for the three values of u, the three values of x which are the roots of the original cubic are known from the relation $u = x + 3$, the number 3 being added to each value of u to give the corresponding x.

Since there are three values of u and three values of x these are denoted by u_1, u_2, u_3 and x_1, x_3, x_3, or for brevity, x_j with $j = 1, 2, 3$. The small figures or letters are not exponents and have nothing to do with powers of u or x, and neither are they coefficients. They are simply numbers or "tags" to distinguish different ones of the values of u or x and are called *subscripts*.

Roots of the Reduced Cubic

In order to solve the complete cubic equation it is first necessary to find the roots of the reduced cubic. The solution of this is itself somewhat complicated and we will not give a full and detailed explanation of the solution $_{□...□}$, but simply an outline of the method and procedure. The solution is found by formula as in the case of the quadratic, but the forms of the formulas are not quite so simple.

The roots of the reduced cubic

$$u^3 + pu + q = 0 \qquad (2.5)$$

with the coefficient p and the constant term q are found as follows:

First, using the values of p and q from the equation, calculate the number

$$Q = \sqrt{\left(\frac{p}{3}\right)^3 + \left(\frac{q}{2}\right)^2}. \qquad (2.6)$$

Using this value of Q, calculate next the two numbers

$$A = \sqrt[3]{Q - \frac{1}{2}q}, \quad B = -\sqrt[3]{Q + \frac{1}{2}q}. \tag{2.7}$$

When the values of A and B are found, the three roots of the reduced cubic 2.5 are then

$$u_1 = A + B$$

$$u_2 = -\frac{1}{2}u_1 + \frac{1}{2}\sqrt{3}(A - B)i$$

$$u_3 = -\frac{1}{2}u_1 - \frac{1}{2}\sqrt{3}(A - B)i,$$

where $i = \sqrt{-1}$.

❧

When the three roots of the reduced cubic have been found, the roots of the original cubic equation are $x_j = u_j - a/3$, where a is one of the original coefficients and $j = 1, 2, 3$ in succession.

❧

As an illustration of the use of this method, we give here the solution of the equation 2.3 already considered above.

(1) The equation in the standard form is

$$x^3 - 9x^2 + 9x - 8 = 0.$$

(2) In this we substitute $x = u - (a/3) = u + 3$ to suppress the second power term. The reduction has already been carried out and the reduced cubic is

$$u^3 - 18u - 35 = 0.$$

(3) In this equation $p = -18$, $q = -35$. Therefore,

$$Q = \sqrt{\left(\frac{-18}{3}\right)^3 + \left(\frac{-35}{2}\right)^2} = \sqrt{(-6)^3 + \left(\frac{35}{2}\right)^2}$$

$$= \sqrt{-216 + \frac{1225}{4}}$$

$$\therefore Q = \sqrt{\frac{361}{4}} = \frac{19}{2}.$$

(4) Using $Q = 19/2$ and $\frac{1}{2}q = -(35/2)$ we find

$$A = \sqrt[3]{\frac{19}{2} - \left(-\frac{35}{2}\right)} = \sqrt[3]{\frac{51}{2}} = \sqrt[3]{27} = 3,$$

$$B = -\sqrt[3]{\frac{19}{2} + \left(-\frac{35}{2}\right)} = -\sqrt[3]{-\frac{16}{2}} = -\sqrt[3]{-8} = 2.$$

(5) The roots of the reduced cubic are then

$$u_1 = 3 + 2 = 5,$$

$$u_2 = -\frac{5}{2} + \frac{\sqrt{3}}{2}(3 - 2)i = -\frac{5}{2} + \frac{3}{2}i,$$

$$u_3 = -\frac{5}{2} - \frac{\sqrt{3}}{2}(3 - 2)i = -\frac{5}{2} - \frac{\sqrt{3}}{2}i.$$

(6) The roots of the cubic equation 2.3 are then

$$x_1 = u_1 + 3 = 8,$$

$$x_2 = u_2 + 3 = \frac{1}{2} + \frac{1}{2}\sqrt{3}i = \frac{1}{2}(1 + i\sqrt{3}),$$

$$x_3 = u_3 + 3 = \frac{1}{2} - \frac{1}{2}\sqrt{3}i = \frac{1}{2}(1 - i\sqrt{3}).$$

The cubic equation $x^3 - 9x^2 + 9x - 8 = 0$, therefore, has one real root, $x = 8$, and two complex° roots $x = \frac{1}{2}(1 + i\sqrt{3})$ and $x = \frac{1}{2}(1 - i\sqrt{3})$.

By the method just illustrated, any cubic equation can be solved, be the roots all real or one real and two complex. If the equation is a complete cubic, the entire procedure is necessary. If, however, the equation does not contain the square of the unknown, the steps 1, 2 and 6 are not necessary.

In any case, if the coefficient of the cube of the unknown is different from 1, the equation must first be divided by that coefficient in order to put the equation in the form 2.1.

Note

complex: a number which involves a multiple of i, the square root of -1.

∾ 3 ∾

"A Goodly Struggle": Problems, Puzzles, and Challenges

"BECAUSE IT'S THERE" IS A GOOD REASON FOR CLIMBING A MOUNTAIN—SO they say—and it seems a good reason for doing a mathematical problem, too. Mathematicians involved in new research had set each other challenges and gone head to head in competitions since at least the sixteenth century, but competitive puzzling at an amateur level boomed early in the eighteenth century with *The Ladies' Diary*, the source of our first extract.

The *Diary*, like *The Athenian Mercury* which we will meet in Chapters 5 and 10, was one of the "new media" of its day, offering readers a chance to become writers, so to speak: to see their mathematics in print, and presumably to become still more enthusiastic buyers of the publication as a result. It was, when it worked, a genuinely interactive mode of publication that seems to have caught the imagination of its middle-class British target audience, and the *Diary* ran for well over a century.

Part of its success was surely its strategy of aiming mathematics at a new audience—women—and the genre of mathematical puzzle writing continued to be marked by a desire to seek out new audiences and to draw new groups of people into mathematical activity. This restless search for new ways to bring mathematics to interested people—and new people to bring it to—is on show in this chapter. Young girls and schoolboys were among the audiences of our later extracts, while in the twentieth century the competitive puzzle for schoolchildren became institutionalized in mathematical olympiads. All of this made for tremendous variation in the difficulty of the problems that were published. The simple conundrums of *The Girl's Own Book* (1835) and of Hirschberg's charming *Can You Solve It?* (1926) bear witness to the fact that, in mathematics more than in most subjects, one person's triviality is another's "goodly struggle."

The Ladies' Diary

1798

A long-running publishing phenomenon of eighteenth-century Britain was *The Ladies' Diary*, an almanac aimed at women which contained, among other things, a section of mathematical puzzles. The puzzles became very popular, attracting solutions from all over England and beyond (some of them from women). The *Diary* ran from 1704 to 1841, generating a number of imitators (the inevitable *Gentleman's Diary*, for one) and spin-off publications (*The Diarian Miscellany*). Here we present selections from near the end of the eighteenth century, when the editor was the well-known mathematician and mathematical educator Charles Hutton, whose textbooks were the source for some of the questions asked.

Charles Hutton (1737–1823), *The Ladies' Diary: or Woman's Almanack ⌊ . . . ⌋ Containing New Improvements in Arts and Sciences, And many Entertaining Particulars: Designed for the Use and Diversion of the Fair-Sex* (London, 1798), pp. 46–48; (London, 1799), pp. 33–34, 43–44.

New Questions

I. by Mr. John Hawkes, of Finedon

What two numbers are those whose product, difference of their squares, and the ratio or quotient of their cubes, are all equal to each other?

II. by Miss Sarah Cowen

Asking lately the contents of a particular field, in form of a trapezium, I was answered that the contents were forgotten, but the dimensions in chains° were as in the annexed figure (3.1); but as I cannot from hence compute the contents, shall be obliged to my friend, Lady Di⌊ary⌋, to do it for me.

III. by Mr W⌊illia⌋m Newby, Barningham

A gentleman has a field in form of a trapezoid, the area of which is two acres, three roods;° the longest side, or base, *AB*, which is one of the two parallel sides, is 1432 links; also the angle at *A* is 14° 17′, and that at *B*

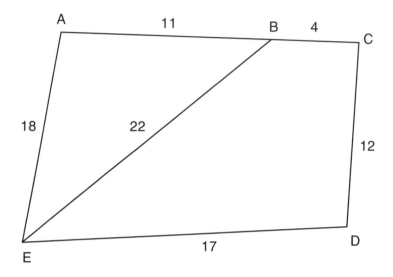

Figure 3.1. Sarah Cowen's field, in the form of a trapezium.

54° 18′. It is required to find what the four sides of the field will cost fencing, at 6 pence a rod.

V. by Mr Henry Armstrong, Bewcastle

There is a vessel in the form of a frustum of a cone,° standing on its lesser base, whose ₊volume₊ is 8.67 feet, ₊its₊ depth 21 inches, its greater base diameter to that of the lesser, as 7 to 5. Into ₊this₊ a globe had accidentally been put, whose ₊volume₊ was $2\frac{1}{2}$ times the measure of its surface. Required, the lineal diameter of the above vessel and globe, and how many gallons of wine would be required just to cover the latter within the former.

VI. by Mr James Sparrow, of Norwich

In a plane triangle, given the angle at the vertex 60°, the length of the line bisecting it and dividing the base into two parts ₊in the ratio₊ 5 to 4, equal to 16, to find the sides and area.

VII. by Mr W₊illia₊m Burden, of Acaster Malbis

How long will the sun be in rising out of the horizon, the 12th of October, 1798, to those inhabitants who have the duration of the twilight at that time the shortest possible?

VIII. by Mr T. Molineux, Macclesfield, Cheshire

Required, a general theorem for determining, by means of a rain gauge, the height of water which falls upon the ground; the weight of a cubic foot of water, the area of the aperture of the rain-gauge, and the weight of the water caught in the vessel being given.

X. by James Glenie, Esq.

On a given ˌstraightˌ line as a base,° to constitute a triangle, such that the ratio of the square on the other two sides, to the ratio of the sum of their cubes to the difference of their cubes, shall be equal to the ratio of 171 to 140, whilst the area of the triangle has to the square of the base a given ratio, suppose that of 1 to 2.

XI. by Mr Thoˌmasˌ Coulthend, of Frosterly

Being in the head of a dale, about seven miles from Kendal, remarkable for the height of the hills on each side, I observed that the road on which I was going intersected at right angles a line joining the middle of the bases of two amazing high, rocky hills; the angle of elevation of the one was 50°, and of the other 60°, taken at the point of intersection. I then proceeded forward in the same direction, up a gentle declivity, rising at an angle of 10°, for the space of 200 yards, and then the angle of altitude of the first was 48° 10′, and that of the other 58°. I desire to know the perpendicular height of the hills from the bottom of the dale, the distance between their summits, and how far I was from the top of each at both stations.

XII. by Mr John Rutherford, Schoolmaster, Lanchester

On Lammas day,° in latitude 54° 40′ north, I observed a tree, which I knew to be 20 yards in height, to cast its shadow along the declivity of a hill. Now, admitting the declivity to be a plane of an indefinite length, and to incline regularly from the east towards the west at an angle of 20°, also the tree to stand perpendicular to the horizon, and exactly at the top of the said declivity, I desire to know how long the shadow was, supposing the time to be an hour before noon?

XIII. by Mr John Ryley, of Leeds

I should like to see a scientific solution to question 69, page 164, of Dr Hutton's *Conic Sections and Select Exercises*, which is this: "If a glass tube, 36 inches long, closed at top, be sunk perpendicularly into water, till its lower or open end be 30 inches below the surface of the water, how high will the water rise within the tube, the quicksilver in the common barometer at the same time standing at $29\frac{1}{2}$ inches?" The answer being only there given, but not the solution.

XIV. by Mr Thomas Milner, Lartington Free School

I have seen a sheep leap from a bridge very high, into water, and swim out. Now, if a globe, whose weight is 112 pounds, and one foot in diameter, fall from an eminence ten yards high, how deep must the water be, just to destroy all the globe's velocity, supposing the density of air, water, and the globe to be in the ratio $1\frac{1}{5}$: 1000 : 10,000, respectively.

XV. or Prize Question, by Amicus

To construct a triangle, having two of the sides given, and such that, a perpendicular to the third side being drawn from its opposite angle, that perpendicular shall be the height of a prism whose base is the triangle, and whose volume is as large as possible.

Answers

I. Question, answered by Mr William Davis, Schoolmaster, of Crowan

Put x = the greater number, and y = the less. Then $xy = x^2 - y^2$ and $xy = x^3 \div y^3$, or $x^2 = y^4$, or $x = y^2$; then by substitution, etc., we have $y^2 - y = 1$. By completing the square, etc., we find $y = \frac{1}{2} + \sqrt{1\frac{1}{4}} = 1.61803$. Consequently $x = 2.61803$.

The Same, by Mr John Eadon, Junior, Sheffield

Let x = the greater, and y = the less number. Then, by the question, $xy = x^2 - y^2$, and $xy = x^3 \div y^3$; therefore $y^4 x = x^3$, and $y^4 = x^2$, and $y^2 = x$. Put y^2 for x in the first equation, and we get $y^3 = y^4 - y^2$, or $y^2 - y = 1$. Hence $y = \frac{1}{2} + \frac{1}{2}\sqrt{5}$, and then $x = y^2 = 1\frac{1}{2} + \frac{1}{2}\sqrt{5}$, which are the two numbers sought.

For proof: $xy = 2 + \sqrt{5}$, and $x^2 - y^2 = 2 + \sqrt{5}$, and $x^3 \div y^3 = 2 + \sqrt{5}$.

The same, by Mr John Ramsay, London

Suppose x the greater number, and y the less. By the question, $xy = x^2 - y^2 = x^3 \div y^3$. By equating the two first quantities is got $x = \frac{1}{2}y \times (1 \pm \sqrt{5})$, and by equating the first and third $x = y^2$; hence $y = \frac{1}{2} \pm \frac{1}{2}\sqrt{5} = 1.618$, etc., or -0.618, etc., and $x = y^2 = \frac{3}{2} \pm \frac{1}{2}\sqrt{5} = 2.618$, etc., or 0.382, etc.

II Question, answered by Mr Thomas Coultherd, Frosterly

Let CE be drawn, and produce CA to F, letting fall the perpendicular EF. Then $AF = (BE^2 - BA^2 - AE^2) \div 2AB = 2.7727$, and $\sqrt{AE^2 - AF^2} = \sqrt{EF^2} = 17.9125$; also $\sqrt{CF^2 + EF^2} = CE = 24.5394$, and $\frac{1}{2}AC \times FE = 134.3437$, the area of the triangle ACE. Again, in the triangle CDE, having the three sides given, by a like process is easily found the perpendicular $GE = 15.469589$, and thence $\frac{1}{2}CD \times EG = 92.8175$, the area. Consequently the sum of these two areas gives 227.162 square chains, or 22 Acres, 2 Roods, $34\frac{1}{2}$ Perches for the area required.

The same, by Mr J. Gee, Elswick, near Newcastle

In the triangle ABE, the three sides are given, so we can find the angle $A = 95° \ 39'$. Hence, if the diagonal CE be drawn, we shall have two sides and the included angle of the triangle ACE, with which we can find the said diagonal $= 24.539$ chains. Then in each of the triangles ACE, DCE, the three sides are known, whence the sum of their areas is easily found $= 22$ Acres, 2 Roods, 35 Perches $=$ the content required.

The same, by Mr Richard Oliver, Assistant to the Rev. Mr Cursham, Sutton, near Mansfield

In the triangle ABE, all the three sides are given, with which we can find the angle $ABE = 54° \ 30'$, the supplement of which is $125° \ 30' =$ the angle EBC. If CE be drawn, we then have the sides EB, BC, and the included angle, whence CE is easily found $= 24.5$. Hence we have the sides of all the triangles AEB, EBC, ECD, from which ⌐...⌐ their areas may be found, the sum of which comes out 22 Acres, 3 Roods nearly.

XIII. answered by Mr J. Gough, Kendal

The density of the air is ⌐proportional to⌐ its spring,° which in the open tube is equal to a column of mercury of the same base and $29\frac{1}{2}$ inches high; but in the immersed tube this weight is increased by a column of water $3 - x$ inches high, x denoting the height of the water in the tube. But $13600 : 1000 = 20 - x : 2.205 - 0.0735x =$ a column of mercury of the same weight, and the whole pressure $= 29.5 + 2.205 - 0.0735x = 31.705 - 0.0735x$. But when the matter is given, the magnitude is inversely ⌐proportional to⌐ the density, or pressure in the present case. Therefore $31.705 - 0.0753x : 29.5 = 36 : 36 - x$; hence $x^2 - 467.36x = -1080$, and $x = 2.33$ inches, as required.

The same, by Miss Maria Middleton, Eden, near Durham

Let $l = 36$ inches the length of the tube, $b = 30$ inches the part immersed, $x =$ the height of water in the tube, and $f = 413$ inches, the height of a column of water equal to the pressure of the atmosphere, when the quicksilver stands at $29\frac{1}{2}$ inches. Then, since the spaces occupied by the same quantity of air are reciprocally as the compressing forces, it will be as $l - x : l = f : \frac{lf}{l-x} =$ force of the air in $l - x$; hence $\frac{lf}{l-x} + x = b + f$, and $x = 2.2654115$ inches.

Notes

chains: a unit of length equal to 66 feet or 100 links.

roods: a unit of area equal to a square perch, $272\frac{1}{4}$ square feet.

frustum of a cone: a cone whose point has been cut off, leaving a figure with two parallel circular faces of different sizes, and one curved face.

On a given straight line ...: this obscure question asks the reader to construct a triangle with sides of lengths a, b, c, where a is given and b and c are to be found, subject to the constraint that the triangle's area should equal $2a^2$ and

$$\frac{b^2/c^2}{(b^3 + c^3)/(b^3 - c^3)} = \frac{171}{140}.$$

Approached using algebra the problem is intractable, but it is possible to solve it quite elegantly using a geometric approach.

Lammas day: 1 August.

spring: pressure.

The Girl's Own Book

Lydia Marie Child, 1835

First published in Boston in 1834 by Lydia Marie Child—also known as a writer of fiction and an activist for women's rights and the abolition of slavery—*The Girl's Own Book* describes a whole range of pastimes and activities, moved by the thought that "every mind should seek to improve itself to the utmost." It went through many editions on both sides of the Atlantic, and a recent reprinting has proved a minor hit, too. The arithmetical puzzles it contains are pitched at the very lowest level of mathematical knowledge, and there is no denying they seem disappointingly trivial: compare those in *The Boy's Own Magazine*, which follow.

Lydia Marie Child (1802–1880), *The Girl's Own Book* (London, 1835), pp. 169–171, 179.

Arithmetical Puzzles

1. How can you take away one from nineteen, and have twenty remain?

2. What is the difference between twice twenty-five and twice five and twenty?

3. If you can buy a herring and a half for three halfpence, how many herrings can you buy for eleven pence?

4. A and B made a bet concerning which could eat the most eggs. A ate ninety-nine; B ate one hundred, and won. How many more did B eat than A?

5. Place four nines together so as to make exactly one hundred.

In the same way four may be made from three threes, three may be made from three twos, etc.

6. If a person hold in his hands a piece of silver, and a piece of gold, you can ascertain in which hand is the silver, and in which the gold, by the following simple process. The gold must be named some *even* number, say *eight*; the silver must be named an *odd* number, say *three*. Then tell the

person to multiply the number in his right hand by an even number, and that in his left hand by an odd number, and make known the amount of the two added together. If the whole sum be odd, the gold is in his right hand; if it be even, the silver is in the right hand. For the sake of concealing the artifice better, you need not know the amount of the product, but simply ask if it can be halved without a remainder; if it can, the sum is, of course, an even one.

7. The figure 9 has one remarkable characteristic, which belongs to no other number. Multiply it by any figure you will, the product added together will still be nine. Thus, twice 9 are 18; 8 and 1 are 9. Three times 9 are 27; 7 and 2 are 9. Eight times 9 are 72; 7 and 2 are nine, etc.

If you multiply it by any figures larger than 12, the result will differ only by there being a *plurality* of nines.

8. When first the marriage knot was tied
    Between my wife and me,
My age exceeded hers as much
    As three times three does three.
But when ten years and half ten years
    We man and wife had been,
Her age approached as near to mine.
    As eight is to sixteen.
Question. How old were they when they married?

9. A room with four corners had a cat in each corner: three cats before each cat, and a cat on every cat's tail. How many cats were there?

10. If you cut thirty yards of cloth into one yard pieces, and cut one yard every day, how long will it take you?

11. How can you show that seven is the half of twelve?

Keys to Puzzles, Conundrums, etc.

1. XIX. XX.

2. Twice twenty-five is fifty; twice five, and twenty, is thirty.

3. If a herring and a half are three halfpence, of course each herring is a penny a piece.

4. Those who hear you will think you say one.

5. $.99 + (9 \div 9).$
8. The bride was 15, and the bridegroom 45.
9. Four cats.
10. Twenty-nine days.
11. XII.
12. Draw a line through the middle of XII, the upper half will be VII.

The Boy's Own Magazine

1855

Beginning in 1855, *The Boy's Own Magazine*, "An Illustrated Journal of Fact, Fiction, History and Adventure," contained a mix of stories, recreations, and nonfiction somewhat in the vein of the later and better known *Boy's Own Paper* (or for that matter *The Dangerous Book for Boys* of 2006). The mathematical questions presented here illustrate not just the sturdy manliness of the publication (rockets and all that) but also the (albeit modest) level of mathematical sophistication expected of its readers. Solutions were not given.

"F.L.J., B.A.," "Mathematical Questions," *The Boy's Own Magazine*, issue 11 (London, 1855), p. 352.

1. *C* spent two 11ths of his life as a child at home, three years more than twice that time as a boy at school, three 11ths as a student at home or at college, and four 55ths in a public appointment. How long was he a schoolboy?

2. At 25 years of age, *C* was 4lbs. 8oz. more than 18 times the average weight of a newly-born infant, but 9lbs. 8oz. added made him just 20 times as heavy. What was his weight?

3. *T*'s expenditure fell short of his income for the first half of the present year by a fifth, but having since increased, by change of residence, he finds that he will spend .0143̇ more than his income for the second half, and have £9, 5 shillings and 8 pence left at the year's end. Find his yearly income.

4. *A* walked from Bedford to Cardington in three-quarters of an hour, and, at a speed two-thirds of a mile per hour less, took as long to reach the top of Hillfoot Hill, after which he reached Warden in an hour, at a rate one-sixth of a mile per hour slower still. Had he walked uniformly

as at first, he would have reached Warden at 24 minutes before 1, but at his last rate not till 14 minutes after. What is the distance from Bedford to Cardington and Warden?

5. *B* noticed that a skyrocket at its highest point of ascent formed an angle 45° of elevation with the horizon. He moved 80 yards nearer to the rocket-post, when another rocket, which mounted as high as the former, made an angle of elevation of 60°. How high did the rockets ascend?

6. *L* saw a boy fishing on the opposite side of a river in a direction forming an angle of 60° with the bank; the boy then moved 42 yards farther down the river, when the angle of his direction with either bank had decreased to 30°. Find the breadth of the river.

7. Standing 5.05 feet from a wall directly facing the sun when the altitude of the latter was 30°, I noticed that my shadow on the wall was just half of my real height. What is my height?

8. From Sandown Castle, Kent (July, '54) *F* noticed the *Hannibal*, man-of-war, and the *Prince*, transport, anchored in the Downs,° in directions forming angles with the direction of Walmer Castle, the *Prince* of 60°, the *Hannibal* of 50°; from Walmer, however, 16 furlongs from Sandown Castle, the direction of the latter made with that of the *Hannibal* an angle of 70°, with that of the *Prince* of 40°. How far from the *Prince* was the *Hannibal*?

Note

The Downs: an anchorage on the southeast coast of England. *Man-of-war* and *transport* are in this context types of ships.

"The Analyst"

1874

Beginning in 1874 and continuing as *Annals of Mathematics* from 1884 onward, *The Analyst* appeared monthly, published in Des Moines, Iowa, and was intended as "a suitable medium of communication between a large class of investigators and students in science, comprising the various grades from the students in our high schools and colleges to the college professor." It carried a range of mathematical articles, both pure and applied, and a regular series of mathematical problems of varying difficulty: on the whole they seem harder than those in *The Ladies' Diary* and possibly easier than the *Mathematical Challenges* in the extract after the next. Those given here appeared in the very first issue.

The Analyst: A monthly journal of pure and applied mathematics, vol. 1, no. 1 (Des Moines, 1874), p. 15.

1. Find the value of x and y in the following equations:

$$a^2x^4 + b^2y^4 = a^2b^2(x + y)^2;$$

$$a^2x^2 + b^2y^2 = a^2b^2.$$

—Communicated by U. Jesse Knisely, President and Professor of Mathematics in Luther College, Newcomerstown, Ohio.

2. Let a regular polygon of 14 sides be described, each of whose equal sides shall be *one*. Then will the radius of its circumscribing circle, which put $= t$, be more than *two* and less than *three*. Put $r = 2 + x$; then is x a positive quantity less than *one*. Let another regular polygon of half the number of sides (7) be inscribed in a circle whose radius is *one*, and determine one of its equal sides in functions of x expressed in its simplest form.

3. If a line make an angle of $40°$ with a fixed plane, and a plane embracing this line be perpendicular to the fixed plane, how many degrees from its first position must the plane embracing the line revolve in order that it may make an angle of $45°$ with the fixed plane?
—Communicated by Prof. A. Schuyler, Berea, Ohio.

4. A cask containing a gallons of wine stands on another containing a gallons of water; they are connected by a pipe through which, when open, the wine can escape into the lower cask at the rate of c gallons per minute, and through a pipe in the lower cask the mixture can escape at the same rate; also, water can be let in through a pipe on the top of the upper cask at a like rate. If all the pipes be opened at the same instant, how much wine will be in the lower cask at the end of t minutes, supposing the fluids to mingle perfectly?
—Communicated by Artemas Martin, Mathematical Editor of *Schoolday Magazine*, Erie, Pennsylvania.

Can You Solve It?

Arthur Hirschberg, 1926

This collection of puzzles, which ranges across mathematics, natural science, and word games of various kinds, promised the reader "a goodly struggle" and that "feeling of intense pleasure which is aroused by the accomplishment of any difficult intellectual task," although many of the puzzles are in fact not all that difficult. The author, Arthur Hirschberg, is otherwise completely unknown.

Arthur Hirschberg (dates unknown), *Can You Solve It? A Book of Puzzles and Problems* (London, 1926), pp. 167–169, 277–278.

Puzzles

1100

Two clerks, *A* and *B*, were hired in an office at the same time. *A*'s salary was £1,000 a year with an increase of £200 a year. *B*'s salary was also to commence at £1,000 a year, but his increase was £50 every half-year. In each case payments were made half-yearly. Which clerk received the better offer, and why?

1101

If five cats catch five rats in five minutes, how many cats will it require to catch one hundred rats in one hundred minutes?

1102

A swimming pool which contains 1000 cubic yards of water can be filled by either of two inlet pipes in three and four hours respectively, and can be emptied by either of two outlet pipes in five and six hours respectively. How long does it take to fill the tank if both inlet pipes and the two outlet pipes are open?

1103

A hymn-board in a church has four grooved rows on which the numbers of the four hymns chosen for the service may be placed. The hymn book contains 700 hymns. The numbers of the hymns chosen are displayed on plates, each carrying one digit. What is the smallest number of plates which can be carried in stock so that the numbers of any four hymns can be displayed? What is the smallest number, if an inverted six can be used for a nine?

1104

A man decides to save a penny on the first day of the month, twopence on the second day, fourpence on the third day, eightpence on the fourth day, and so on—doubling the amount to be saved each day. How much does he save in a month of 31 days?

1105

Two clubs were formed. The first had 15 members and it was agreed upon to dine four at a time until all possible combinations have been formed.

The second club consisted of 21 members and dined in groups of three at a time, until all possible combinations had been formed.

Which group served more dinners, and how many more?

1106

In a potato race, each contestant started from a basket, ran 3 yards to the first potato, returned and deposited the potato in the basket. He repeated this with each of the other potatoes, but each potato was 3 yards farther from the basket. If each contestant picked up 24 potatoes in this manner, what was the total distance he ran?

Answers

1100

B is better off. His income is always £50 a year more than *A*'s. In the first year *A* received £1,000, and *B* receives £500 for the first half-year, and £550 for the second half-year, or a total of £1,050. In a similar manner it can be shown that there will always be a difference of £50 in *B*'s favour.

1101

The same number—that is, 5 cats.

1102

$4\frac{8}{13}$ hours.

1103

Nine plates, each carrying digits from 1 to 6, inclusive, ⌐and⌐ eight plates, each carrying the digits 0, 7, 8, 9, are needed. This makes a total of 86 plates.

If inverted sixes are used for nines, then twelve sixes must be provided, but no nines are needed. This makes a total of 81 plates.

1104

£2,147,483,647.

1105

The first group had to serve 1,365 dinners, and the second group 1,330 dinners; or, 35 more dinners were served by the first group.

1106

1,800 yards.

Mathematical Challenges

1989

We have seen competitive mathematical puzzles from the eighteenth century in *The Ladies' Diary*, and from the nineteenth century in *The Analyst*. In the twentieth century more formal competitions began to take place: mathematical olympiads have run in various countries, and, since 1959, internationally. All such mathematical competitions rely on a plentiful supply of mathematical problems of a suitable kind: typically they need to be hard yet require little factual knowledge (all the same, the explanation of what a prime number is, in the solution to question 2 in this extract, comes as something of a surprise). Here we present selections from one collection of such material, the *Mathematical Challenges* of the Scottish Mathematical Council. We present two sets of problems and one of solutions, leaving the solutions of the second set as an exercise for the reader.

Mathematical Challenges (Edinburgh, 1989), pp. 30–31, 37, 128–30.

⌐Problems: first set⌐

1. Ten married couples sit down for dinner at a circular table. Each man sits opposite his wife. After the meal, a wit in the company points out that if all the ladies get up and move the same number of places to the right then each one will end up sitting on the knee of some gentleman other than her husband. How were the couples arranged around the table?

2. A printer makes mistakes in setting up type and prints $5^4 2^3$ as 5423. He makes similar mistakes with another product but this time it does not matter. What four-digit number does he print? (Solutions involving the use of a computer will not be accepted.)

3. Show that the equation

$$x^4 + 131 = 3y^4$$

has no solution in integers x and y.

4. In a certain garden there are several beds each planted with flowers. For any two different varieties of plants in the garden, there is exactly one bed containing both. If any bed is considered, there is exactly one other bed in which none of the same varieties of plants is growing.

Show that

   (i) every bed has at least two varieties,
  (ii) there are at least four varieties in the garden,
 (iii) there are at least six flower beds, and
 (iv) no bed has more than two varieties.

Problems: second set

1. A wine merchant had twelve containers of the following capacities: 13, 15, 16, 18, 19, 21, 31, 40, 50, 51, 81 and 91 litres. One container contained beer for himself, and the rest, which contained wine, were sold intact. All the wine was sold to two customers. The second customer bought twice the amount of wine bought by the first customer.

How much wine did each customer buy?

2. Prove that the only solution in integers of the equation

$$x^2 - 2xy + 2y^2 - 4y^3 = 0$$

is $x = 0$, $y = 0$.

3. In the diagram (Figure 3.2), A, B and C are the centres of three circles each of which touches the other two.

Prove that the perimeter of the triangle ABC is equal to the diameter of the circle with centre C.

4. You are given a red, a white and a blue marble, all of the same weight, and also a red, a white and a blue marble again equal in weight to each other but heavier than the first three. The heavier marbles look identical to the lighter ones. Describe a procedure to find the three heavier marbles in two weighings on a balance scale.

Solutions

1. Number the places 0, 1, 2, ..., 19, starting with 0 for the seat occupied by one of the men and then moving counter-clockwise around the table.

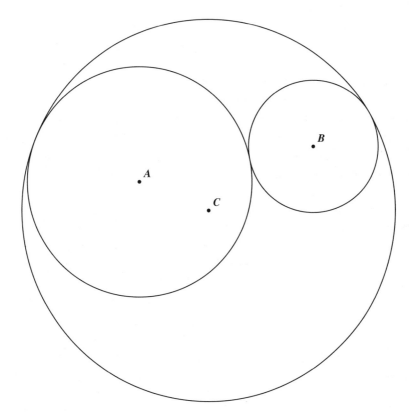

Figure 3.2. Three circles each of which touches the other two.

Then label each seat M or W depending on whether the occupant at the beginning is a man or a woman.

Suppose that all the ladies were to move n places to the right (that is, counter-clockwise around the table). Then $0 < n < 20$. After the move each seat labelled M would have two occupants and each seat labelled W would be empty. Thus n places to the right of a seat labelled W is one labelled M. Furthermore a seat n places to the right of one labelled M could not be occupied by a lady after the move and therefore would be empty, hence it is labelled W. Therefore since 0 is an M, it follows that n is a W, the seat with number equal to the remainder on dividing $2n$ by 20 is an M, that with number equal to the remainder on dividing $3n$ by 20 is a W and so on.

Suppose now that n is odd. Then $10n$ leaves remainder 10 on division by 20. Hence seat 10 is an M. But this cannot be true, since seat 10 is opposite seat 0 and so is occupied by the wife of the man in seat 0. Hence n is even. Next suppose that n is a multiple of 4. Then $5n$ leaves remainder 0 on division by 20. By the statement at the end of the previous paragraph this seat is a W; but we know that it is an M. Hence n is not a multiple of 4.

This shows that the possible values of n are 2, 6, 14 and 18. No matter which of these we take we find that the arrangement must follow the pattern $MMWWMMWW....$

2. A prime is a positive integer which cannot be expressed as the product of two smaller positive integers. For example, 2, 3, 5 and 7 are primes but 4, 6, 8 and 9 are not. Any positive integer greater than 1 is either a prime or can be expressed as a product of primes (with possible repetitions). We shall use this as a basis for an organised search for solutions of the equation $abcd = a^b c^d$.

Since a, c are single digit numbers, the primes which feature when $a^b c^d$ is expressed as a product of primes are all less than 10 and so the possibilities are 2, 3, 5 and 7. Moreover the only number less than 10 involving two different primes in a factorization is $6 = 2 \times 3$ and so $a^b c^d$ can be expressed in one of the following forms:

(i)	2^n	(ii)	3^n	(iii)	5^n	(iv)	7^n
(v)	$2^m 3^n$	(vi)	$2^m 5^n$	(vii)	$2^m 7^n$	(viii)	$3^m 5^n$
(ix)	$3^m 7^n$	(x)	$5^m 7^n$	(xi)	$6^m 5^n$	(xii)	$6^m 7^n$.

Here m and n are positive integers.

We check these possibilities in turn, listing all the four digit numbers of the required form. First we have the powers of a prime.

(i) 1024, 2048, 4096, 8192.
(ii) 2187, 6561.
(iii) 3125.
(iv) 2401.

None of these has the required property. Next we consider case (v), listing first three times a power of 2, then 3^2 times a power of 2 and so on.

(v) 1536, 3072, 6144; 1152, 2304, 4608, 9216; 1728, 3456, 6912; 1296, 2592, 5184, 1944, 3888, 7776; 1458, 2916, 5832; 4374, 8748. Of these, just one number has the required property: $2592 = 2^5 9^2$.

(vi) In this case, each number is divisible by 2 and by 5 and hence is divisible by 10. Therefore if $abcd$ is to equal $a^b c^d$, the digit d is 0; but then $abcd = a^b$, which requires a to be divisible by both 2 and 5, which is not possible.

In the remaining cases, with the exception of (xi), we use a system similar to that followed in (v). Case (xi) is similar to case (vi).

(vii) 1792, 3584, 7168; 1568, 3136, 6272; 1372, 2744, 5488; 4802, 9604.

(viii) 1215, 3645; 2025, 6075; 1125, 3375; 1875, 5625; 9375.

(ix) 1701, 5103; 1323, 3969; 1029, 3087, 9261; 7203.

(x) 1715; 1225, 8575; 6125; 4375.

(xi) 2058; 1764; 1512; 9072.

In all these cases, no number listed has the required property. Hence the only number which has the property is 2592, which is equal to $2^5 9^2$.

3. The last digit of the square of an integer is either 0, 1, 4, 5, 6 or 9. A fourth power is the square of a square, and so its last digit is either 0, 1, 5 or 6. Hence the last digit of $x^4 + 131$ is either 1, 2, 6 or 7 whilst the last digit of $3y^4$ is either 0, 3, 5 or 8. It follows that an integer cannot be both of the form $3y^4$ and of the form $x^4 + 131$. Therefore the equation $x^4 + 131 = 3y^4$ has no integer solutions.

4. There are three rules:

(P) All beds contain flowers.

(Q) For any two varieties of flower there is exactly one bed containing both.

(R) If X is a flower bed there is exactly one bed which contains none of the varieties from bed X.

(i) Suppose there is a bed A which contains just the one variety, α. By rule (R) there is exactly one bed which does not contain α. Call this bed B. B contains flowers (rule (P)). Let β be one of the varieties in B. By rule (Q) there is a bed, C, containing both α and β. Clearly C is not either A or B. Now use rule (R) again to deduce the existence of a bed D containing neither α nor β. Bed D is not the same as bed B. We now know of two beds—B and D—neither of which contains any of the varieties from A. This is contrary to rule (R). Consequently A cannot contain just the one variety.

(ii) Label one of the beds A. In view of (i) A contains at least two varieties, say α, β. By rule (R) there is a bed B containing neither α nor β. Because of (i) B must contain two varieties. Call them γ, δ. We now have four distinct varieties.

(iii) Continuing the argument of (ii) we have bed A which contains α, β but not γ, δ and bed B which contains γ, δ but not α, β. By rule (Q) there is a bed C containing α, γ. If C were to contain β or δ we should contradict the "exactly one" part of rule (Q) applied either to varieties α, β or to varieties γ, δ. So C contains α, γ but not β, δ.

Similarly we can deduce the existence of beds D, E, F such that

(a) D contains α, δ but not β, γ.
(b) E contains β, γ but not α, δ.
(c) F contains β, δ but not α, γ.

These six beds A–F are clearly all different, and so we have established (iii).

(iv) Let A be one of the beds and suppose it contains m varieties. By rule (R) there is a bed B containing no varieties from A. Suppose B contains n varieties. We count the number of beds.

It follows from rules (Q) and (R) that all beds other than A and B contain exactly one variety from A and exactly one variety from B. So there are mn beds other than A and B. So there are $mn + 2$ beds in total.

Let α be a variety from A. It appears once with each variety from B and once with no varieties from B (the latter in bed A). So α appears in $n + 1$ beds.

Let C be a bed other than A or B, and suppose C contains q varieties. By rule (R) there is a bed D containing no varieties from C. D does however contain a variety β from bed A. It follows from the previous paragraph that B appears in $n + 1$ beds. However if we repeat the argument of the previous paragraph but with D, C replacing A, B respectively then we see that β appears in $q + 1$ beds. It follows that $q = n$. This shows that all beds, with the possible exception of A, contain n varieties. Now repeat the argument of the second paragraph with C, D in place of B, A. This will show that there are $n^2 + 2$ beds. So $n^2 + 2 = mn + 2$. Therefore $m = n$ so all beds have n varieties.

Now we count the number of varieties. There are $n^2 + 2$ beds each containing n varieties. Each variety appears in $n + 1$ beds. So the number of varieties is

$$\frac{n(n^2 + 2)}{n + 1}.$$

This has to be an integer.

$n + 1$ and n have no common factor. So if

$$\frac{n(n^2 + 2)}{n + 1}$$

is to be an integer $n + 1$ must divide $n^2 + 2$.

$n^2 + 2 = (n + 1)^2 - 2(n + 1) + 3$. So if $n + 1$ divides $n^2 + 2$ we must have $n + 1$ dividing 3. This will only happen if $n = 2$.

∻ 4 ∻

"Drawyng, Measuring and Proporcion": Geometry and Trigonometry

"Geometry teacheth the drawyng, Measuring and proporcion of figures," wrote Robert Recorde. It also teaches logical thought, and for many years geometry was taught both as a worthwhile study for its own sake and also as a way to discipline the mind and make it fit for other studies.

Unlike arithmetic, geometry did have a Euclid, and the *Elements*, "the greatest [textbook] the world is privileged to possess" (see Thomas Heath in Chapter 7), proved all but irresistible as a model for generations of textbook writers. Straightforward translations of the *Elements* into English number dozens; many and many were the geometry books which translated its text, reproduced its theorems, or imitated its soothing logical manner. All of the extracts in this chapter show the influence of the *Elements* in one way or another.

Like Chapter 2, this chapter presents a short course in basic mathematics, this time running from the definitions of points and lines in 1551 to Napier's rules for spherical triangles in 1956. (This chapter thus approaches the threshold of one of the most important uses of mathematics: navigation on a spherical earth; see Chapter 8 for more.) And, as in Chapter 2, there are changes of style and presentation but also an important second strand in the content of what was taught. This time that second strand is trigonometry, which, particularly from about 1700 onward, provided, in popular presentations, an alternative way of studying triangles based on the use of numerical tables rather than the manipulation of instruments or the mental juggling of propositions and proofs. Trigonometry is a hard subject to present historically since, as I explain in the introduction to the extract from Edward Wells (1714), the basic definitions have subtly shifted in a way that can make a lot of early trigonometry very hard to read. At the same time it often depends on trigonometric tables (canny authors tailored their writings to their own particular tables to increase profits) and means little without them. I have done what I can.

Points and Lines

Robert Recorde, 1551

We have met Robert Recorde before; his arithmetic book features in Chapter 1, and indeed it dominated the textbook market for much over a century. He led the way, too, in the writing of geometry textbooks in English, and here we see him setting out a few first geometrical definitions. His interest in practical instruction rather than the austere Euclidean manner gives his presentation of geometry a very particular—and attractive—flavor.

Robert Recorde, *The pathway to Knowledg, Containing the First Principles of Geometrie, as they may more aptly be applied unto practise, both for the use of instrumentes Geometricall, and astronomicall and also for proiection of plattes in euerye kinde, and therfore much necessary for all sortes of men* (London, 1551), Ai^r–Aii^r.

A Point, or a Prick, is named of Geometricians that small and unsensible shape, which hath in it no parts: that is to say, neither length, breadth, nor depth. But as this exactness of definition is more meet for only Theoric speculation than for practise and outward work (considering that mine intent is to apply all these whole principles to work), I think meeter for this purpose to call "point" or "prick" that small print of pen, pencil, or other instrument, which is not moved, nor drawn from its, first touch, and therefore hath no notable length nor breadth: as this example doth declare: ∴.

Where I have set three pricks, each of them having both length and breadth, though it be but small, and therefore not notable.

Now, of a great number of these pricks is made a Line, as you may perceive by this form ensuing: .... where I have set a number of pricks. So if you, with your pen, will set in more other pricks between every two of these, then will it be a line, as here you may see: ——— and this line is called of Geometricians, Length without breadth.

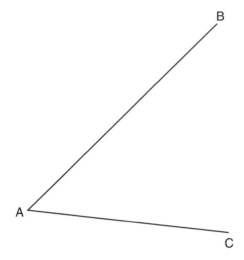

Figure 4.1. Two lines meeting at one notable prick.

But as they, in their theorics (which are only mind-works), do precisely understand these definitions, so it shall be sufficient for those men which seek the use of the same things as sense may duly judge them, and apply to handiworks if they understand them so to be true, that outward sense can find none error therein.

Of lines there be two principal kinds. The one is called a right or straight line, and the other a crooked line.

A Straight line is the shortest that may be drawn between two pricks.

And all other lines, that go not right forth from prick to prick, but boweth any way: such are called Crooked lines ⌊...⌋. So now you must understand, that every line is drawn between two prickes, whereof the one is at the beginning, and the other at the end.

Therefore whensoever you do see any forms of lines to touch at one notable prick, as in this example (Figure 4.1), then shall you not call it one crooked line, but rather two lines, in as much as there is a notable and sensible angle by A which evermore is made by the meeting of two ⌊separate⌋ lines. And likewise shall you judge of this figure (4.2), which is made of two lines, and not of one only.

So that whensoever any such meeting of lines doth happen, the place of their meeting is called an Angle or corner.

Figure 4.2. A figure made of two lines, and not of one only.

Of angles there be three general kinds: a sharp angle, a square angle, and a blunt angle. The square angle, which is commonly named a right corner, is made of two lines meeting together in form of a square. Which two lines, if they be drawn forth in length, will cross one another ⌊at right angles⌋. ⌊...⌋

A sharp angle is so called because it is lesser than is a square angle, and the lines that make it do not open so wide in their departing as in a square corner. And if they be drawn cross, all four ⌊angles formed⌋ will not be equal.

A blunt, or broad, corner is greater than is a square angle, and his lines do part more in sunder than a right angle.

Squares and Triangles

Thomas Rudd, 1650

"Captain Thomas Rudd, chief Engineer to his late Majesty" was one of the foremost British military engineers of his day until the British civil wars–and his Royalist politics–ended his practical career. Thereafter he devoted himself to writing, producing the two-volume account of practical geometry from which this extract is taken, as well as a version of Euclid, and other works. His "practical geometry," like Recorde's, operates at one remove from purely Euclidean methods, but it takes the material in a somewhat different direction. Here we see him explaining various operations to do with the sizes and areas of plane figures.

Thomas Rudd (1583/4–1656), *Practical Geometry, In Two Parts: The First, Shewing how to perform the four Species of Arithmetic, (viz. Addition, Substraction, Multiplication, and Division,) Together with Reduction, and the Rule of Proportion in Figures. The Second, Containing a Hundred Geometricall Questions, With their Solutions and Demonstrations, some of them being performed Arithmetically, and others Geometrically, yet all without the help of Algebra. A Worke very necessary for all Men, but principally for Surveyors of Land, Engineers, and all other Students in the Mathematicks.* (London, 1650), part 1, pp. 3–5; part 2, pp. 1–3.

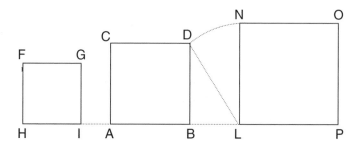

Figure 4.3. Two squares given, to be added.

Two Squares being given to be added, to find their sum

Construction

Let there be to be added two Squares $ABCD$ and $FGHI$, and let one of the sides of the one square, as of AB, be prolonged at pleasure, and let the side of the other square be set upon the prolonged line, beginning from the point of the line prolonged, till the end of the square. As, for example (see Figure 4.3), we prolong the side AB infinitely, towards K, then we set the side FG from the point B to L, and we draw the line DL. I say then that DL is ˎequal toˏ the side of a square containing as much as the two squares given—$ABCD$ and $FGHI$—together.

Demonstration

The Demonstration is manifest by the 47ˎth propositionˏ of the 1ˎst book ofˏ Euclid.° For the square which is made of the side LD is equal to the squares made of the sides LB, BD. But LB is the side of the square $FGHI$, as also the side BD ˎisˏ equal to the square DA. Wherefore the square that is made of the line LD is equal to the squares aforesaid, which was required to be demonstrated.

Question

There is a Triangle, as ABC, whereof the side AB is 13, and AC is 15, and BC is 14 (see Figure 4.4). The Question is, How long is the Perpendicular Line AD?

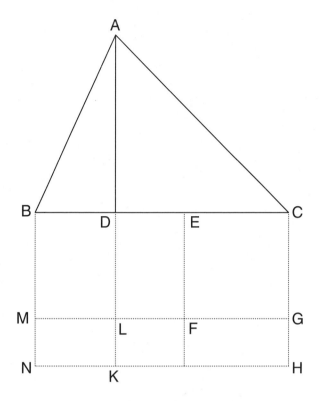

Figure 4.4. How long is *AD*?

To know this, note that if the Square of *AB* be subtracted from the Square of *AC*, there shall then even as much remain, as if the Square of *BD* were taken from the Square of *DC*. (For as much as the Squares of *AD* and *DB* together are equal to the Square of *AB*, and also the Squares of *AD* and *DC* are together equal to the Square of *AC*, by the 47th proposition⌟ of the 1⌞st book⌟ of Euclid.) Now the Square of *AB* is 169, and the Square of *AC* is 225; take therefore that 169 from 225, and there shall ⌞remain⌟ 56 for the Gnomon° *DEFGHKD*, which Gnomon is equal to *GMNH*, by reason that *DEFL* is equal to *LKNM*.

Then is *MGNH* also 56: and seeing the length *NH* or *BC* is known to be 14, therefore divide the same 56 by 14. The Quotient is 4, for *NM* is equal to *DE*; subtract *DE*, 4, from *BC* 14. ⌞There remains⌟ 10 for *BD* and *EC* together, and seeing that *BD* and *EC* are equal, the half of 10 is 5 for *BD*,

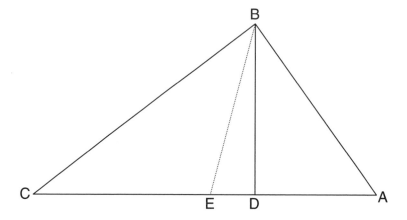

Figure 4.5. Rudd's second triangle: how long is *BD*?

and (by the aforesaid 47th proposition of the 1st book of Euclid) is *AD* known by subtracting the Square of *BD*, 25, from the Square of *AB*, 169. There remains the Square of *AB*, 144, whose Square Root is 12 for *AD*, which was required.

Question

In this Triangle *ABC* (see Figure 4.5), the angle *B* is a right angle, whose perpendicular *DB* is unknown, but the parts of the base where the same shall fall are known, for *AD* is 18, and *DC* is 32.

The Question is: How much is the perpendicular line *BD*?

Seeing that *AD* is 18, and *DC* 32, it followeth that the whole base, *AC*, is 50, whose half is 25, for *AE* or *EC* (or for *EB*, because the angle *B* is a right angle). Therefore, take *AD*, 18, from *CD*, 32; there remains 14, whose half is 7 (or take 18 from 25, the half of 50). So we have 7 for the difference *DE*. And seeing now that *BE* can be known, for that *AE*, *EC*, and *EB* are equal (by reason that *B* is a right angle), that is, each of them is 25—therefore (by the 47th proposition of the 1st book of Euclid), can *BD* be also known, by subtracting the square of *ED*, 49, from the square of *BE*, 625. There remains the square of *BD*, 576, whose square Root is 24 for *BD*, the perpendicular line required.

Note

The 47th proposition ...: Pythagoras's Theorem.
Gnomon: "the part of a parallelogram which remains after a similar parallelogram is taken away from one of its corners." (*OED*).

Pythagoras's Theorem

Edmund Scarburgh, 1705

For generations of schoolchildren, geometry meant Euclid, and it meant in particular a handful of key theorems, of which we present one of the most famous here: Pythagoras's Theorem. Euclid had first been translated into English in 1570 (there is a section of the preface to that translation in Chapter 10); many other translations followed, often differing little in their choice of words, diagrams, or notation but showing more variation in the annotations and supplements they provided. Scarburgh, as we see here, attempts to give some hint of how the theorem might have been first discovered: compare Joseph Fenn's imagined "first analysts" in Chapter 3.

Edmund Scarburgh, *The English Euclide, being The First Six Elements of Geometry, Translated out of the Greek, with Annotations and useful Supplements* (Oxford, 1705), pp. 108–109.

In a Right-angled Triangle, the Square of the side subtending the Right angle is equal to the Squares of the sides containing the Right angle

Let the Right-angled Triangle be *ABC*, having the Right angle *BAC* (see Figure 4.6). I say that the square of *BC* is equal to the squares of *BA*, *AC*.

For on *BC*, let be described the square *BDEC*, and on *AB*, *AC*, the squares *GB*, *HC*, and by *A* let *AL* be drawn parallel to either of the lines *BD*, *CE*, and let be joined *AD*, *FC*.

Now forasmuch as each of the angles *BAC*, *BAG*, is a Right angle, and to the straight line *BA*, and to a point in the same *A*, the two straight lines *AC*, *AG*, not lying the same way, make the consequent angles equal to two Right, therefore *CA* is ⸤parallel⸥ to *AG*. By the same reason also *AB* is ⸤parallel⸥ to *AH*. And because the angle *DBC* is equal to the angle *FBA*, for each is a Right angle, let the angle *ABC* be added in common, therefore

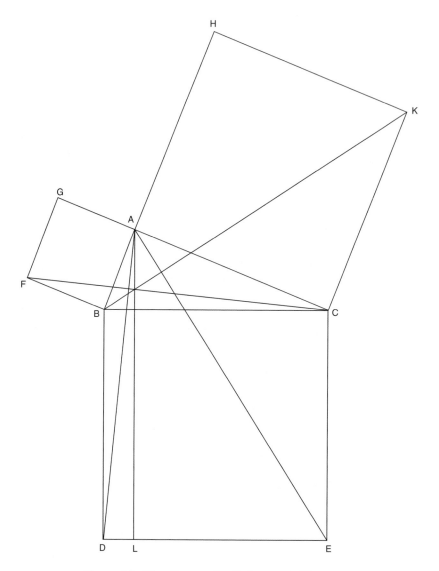

Figure 4.6. The diagram for Pythagoras's Theorem.

the whole angle *DBA* is equal to the whole angle *FBC*. And because the two lines *DB*, *BA*, are equal to the two lines *CB*, *BF*, each to each, and the angle *DBA* is equal to the angle *FBC*, therefore the base *AD* is equal to the base *FC*, and the Triangle *ABD* is equal to the Triangle *FBC*.

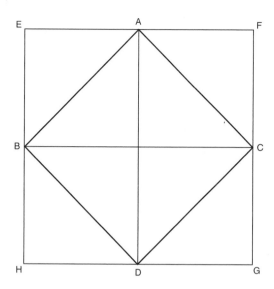

Figure 4.7. How Pythagoras's Theorem might have been first conceived. (Scarburgh, p. 109.)

Now the Parallelogram *BL* is double of the Triangle *ABD*, for they have the same base *BD*, and are in the same parallels *DB*, *AL* (Proposition 41). Also the square *GB* is double of the Triangle *FBC*, for they have the same base *FB*, and are in the same parallels *FB*, *GC*. Now the doubles of equals are equal to one another; therefore the Parallelogram *BL* is equal to the square *GB*.

In like manner, *AE*, *BK* being joined, may be proved that the Parallelogram *CL* is equal to the square *HC*; therefore the whole square *BDEC* is equal to the two squares *GB*, *HC*. And the square *BDEC* is described on *BC*, and *GB*, *HC* on *BA*, *AC*; wherefore the square of the side *BC* is equal to the squares of the sides *BA*, *AC*.

Therefore in Right-angled Triangles, the square of the side subtending the Right angle is equal to the squares of the sides containing the Right angle. Which was to be demonstrated.

This Proposition, among Geometricians most famous, is said to have been found out by *Pythagoras*, and the Invention publicly celebrated with a Sacrifice to the Muses. Yet the hint from whence the discovery of this Truth might first arise, seems to be very obvious.

For in this Figure (4.7) the square *EFGH* is apparently double of the square *ABDC*. But *EFGH* is described on *EF*, which is equal to

BC, the side subtending the Right angle *BAC* of the ˌisoscelesˌ Triangle *ABC*; and the square *ABDC* is described on either of the sides *AB, AC,* containing the Right angle *BAC,* of the same ˌisoscelesˌ Triangle *ABC.* It is therefore hereupon very reasonable to conceive that the same property might likewise belong to Scalene Right-angled Triangles, and give the occasion of a farther enquiry into this matter.

Thus Geometricians often happen to discover a Truth before they have framed a legitimate demonstration of it, and find out their Propositions one way (which they usually conceal) but prove them in another. We have an Example of this kind in the Remains of *Archimedes,* who shows how first he found the Quadrature of a Parabola° Mechanically, as he calls it, and afterwards gives a Geometrical demonstration.

Note

Quadrature of a parabola: the area under a parabola.

Trigonometrical Definitions

Edward Wells, 1714

It is very difficult to find approachable introductions to trigonometry from this period because—as this extract shows—the terms "sine," "cosine," and so on, were used for *lengths* in a particular construction involving a circle: not, as now, for *ratios* of lengths in a right-angled triangle. This way of using the terms remained in use until the second half of the nineteenth century, and it is not immediately obvious that—as is in fact the case—it is numerically equivalent to the modern way.

Edward Wells's training was in divinity and his profession that of an Anglican clergyman, but he had an abiding interest in education and put together textbooks ("more voluminous than distinguished," according to a recent biographer) on subjects from mathematics to geography, history, and religion. His trigonometry is, for its period, unusually lucid.

Edward Wells (1667–1727), *The Young Gentleman's Trigonometry, Mechanicks, and Opticks. Containing such Elements of the said Arts or Sciences as are most Useful and Easy to be known.* (London, 1714), pp. 1–5.

The Explication of Trigonometrical Terms

Every Triangle consists in all of seven Parts, *viz.*, three Sides, three Angles, and the Area or Space comprehended by the Sides. Now, although the Word *Trigonometry* does in the Greek Language literally signify *the Measuring of a Triangle*; yet, in its proper Sense, or as it is distinguished from *Geometry*, it does not denote the Measuring of the *Area* of a Triangle, this being deduced from other than what are properly called Trigonometrical Principles, but it denotes only the Measuring, or rather the Art of finding the Measure of, the unknown *Sides and Angles* of a Triangle, by the Help of those that are given or known.

And the actual finding of the unknown Sides and Angles of a Triangle is called, in one Word, the *Solution* or *Resolving* of a Triangle.

The Solution of Triangles is founded on that mutual Proportion which is between the Sides and Angles of any Triangle. And this Proportion is known by finding the Proportion which the ⸢radius⸥ of a Circle has to certain other ⸢straight⸥ Lines applied to the same Circle. And these ⸢straight⸥ Lines are distinguished in general by the Names of *Chords*, *Sines*, *Tangents*, and *Secants*.

A *Chord* is a ⸢straight⸥ Line drawn from one Extremity of any Arc to the other. (As an *Arc* is so called, because it resembles a bent Bow, so a *Chord* is so called, because it resembles the String of a bent Bow. It is otherwise called a *Subtense*, because it *subtends* (i.e. is extended under) the Arc from one End of it to the other.) Thus (see Figure 4.8) *CH* is the Chord both of the lesser Arc *CTH*, and also of the greater Arc *CDH*. And it is to be noted that a Chord of 60 Degrees is always equal to the ⸢radius⸥ of the same Circle.

A *Sine* is either a *Right* or *Versed* Sine. A *right Sine* is one Half, *CA*, of a Chord *CH*, bisected by the Diameter *DT*. Or, ⸢equivalently,⸥ it is a Perpendicular, *CA*, let fall from *C*, one End of an Arc *CT*, ⸢and perpendicular to⸥ a Diameter *DT* drawn from *T*, the other End of the Arc. A *right Sine* is frequently styled simply a *Sine*, and it is to be observed that by the Word *Sine*, put simply or by itself, is always to be understood a *right Sine*.

It is also to be observed that, according to the foregoing Definition of a Sine, the Sine *BK* of an Arc *KT* of 90 Degrees is always equal to the ⸢radius⸥ of the same Circle, and so ⸢is⸥ the greatest Sine that can be ⸢for that circle⸥. Whence it is called the *total* or *whole* Sine.

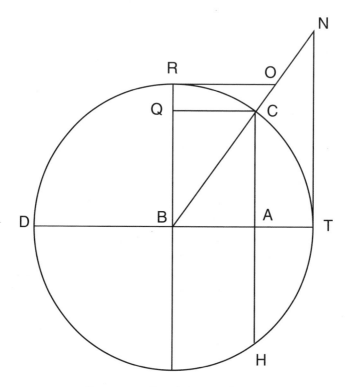

Figure 4.8. Chords, arcs and sines.

A *versed* Sine is the Segment of the ⌐radius ...⌐ between the right Sine and the Arc. Thus, *AT* is the versed Sine of the Arc *CT*, and *AD* the versed Sine of the Arc *CD*.

A *Tangent* and a *Secant* mutually explain one ⌐an⌐other. For if a right Line, *BN*, of an indeterminate Length, be drawn from the Center *B*, through *C* one End of an Arc *CT*, and ⌐if⌐ from *T*, the End of the Diameter *DT*, there be erected a Perpendicular *TN*, extended till it meets with *BE*, ⌐then⌐ the said Perpendicular *TN* will be the Tangent of the Arc *CT*, and the Line *BN* will be the Secant of the same Arc. And it is to be noted that the Tangent of 45 Degrees is always equal to the ⌐radius⌐ of the same Circle; as may be seen (Figure 4.8).

As, when the Arc *CT* is less than 90 Degrees, the Arc *CK* which, being added to the Arc *CT*, does complete or make it up an Arc of 90 Degrees, thence is called the *Complement* of the Arc *CT*; so (see Figure 4.8) the Sine *CQ*, the Tangent *KO*, the Secant *BO*, and the versed Sine *KQ* of the Arc *CK*, are respectively called the *Sine*, the *Tangent*, the *Secant*, and the *versed Sine* of the *Complement* of the Arc *CT*. Or, in short, the *Cosine*, the *Cotangent*, the *Cosecant*, and the *Coversed ˌsineˌ* of the said Arc *CT*.

The Resolution of Triangles

Hugh Worthington, 1780

Hugh Worthington was better known as a dissenting minister, writer, and preacher (his immediately previous publication was entitled *The Progress of Moral Corruption*). The present, unusual, work discusses in some detail various strategies for rapidly computing the lengths and angles of triangles (the "plain" of the title means "plane" as opposed to "spherical"), showing something of the strategies of approximation and "quick fix" that were used in practical mathematical situations in the late eighteenth century, as well as Worthington's own method, which involved the use of a table of what he called "natural radii" for each angle.

Hugh Worthington, Junior (1752–1813), *An essay on the Resolution of Plain Triangles, by Common Arithmetic: with a New and Concise Table, adapted to the Purpose* (London, 1780), pp. 1–5, 16–18.

The doctrine of *right-angled* triangles is generally distributed into *seven* cases, but these are reducible to *four* when the operation does not depend upon the tables. I shall place them in the order most convenient to my purpose.

The FIRST CASE is that of, *the legs being given, to find the hypotenuse*: which is solved by Euclid 1.47, or "the square-root of the sum of the squares of the legs ˌ=ˌ hypotenuse."

The SECOND CASE is, *the hypotenuse and one leg ˌbeingˌ given, to find the other leg.* Here it is evident that "the square-root of the difference of the squares of the hypotenuse and given leg" must be the precise value of the other. Or more easily thus. Since the ˌproductˌ of the sum and difference of any two numbers is equal to the difference of their squares, "multiply the sum of the hypotenuse and one leg by their difference; the square root of that product will be the other leg required."

The THIRD CASE is, *the sides being given, to find the angles,* and the rule is as follows. "Half the longer of the two legs added to the hypotenuse, is always in proportion to 86, as the shorter leg is to its opposite angle."

The FOURTH and LAST CASE is, *the angles and a side being given, to find another side.* For the solution of which problem, the following rule has been proposed by some writers:

"In any right angled triangle, as CAB, let a be the lesser angle in degrees; then

$$\frac{57.3}{a} + \frac{3a}{1000} : 1 = \text{hypotenuse } AC : \text{opposite side } AB, \text{ nearly,}$$

and

$$1 : 1 - \frac{1\frac{1}{2}aa}{10000} = \text{hypotenuse } AC : \text{adjacent side } BC, \text{ nearly.}"$$

Mr. Turner, in his "Plain Trigonometry rendered easy and familiar," has revived the solution first given by Mr. Henry Wilson, in his *Navigation new modelled* (published about 1715), which is to this effect:

"Divide 4 times the square of the complement of the angle whose opposite side is either given or sought, by 300 added to 3 times the said complement. This quotient, added to the said angle, will give you an artificial number, called sometimes the *natural radius,* which will ever bear the same proportion to the hypotenuse as that angle bears to its opposite side.

"In angles under 45 degrees, the *artificial number* may be found easier thus: divide 3 times the square of the angle itself, whose opposite side is given or sought, by 1000; the quotient added to 57.3 (the *radius* of a circle, whose periphery is 360), that sum will be the artificial number required."

It may not be improper to observe, that the latter part of Mr. Wilson's rule is virtually the same with the first proportion, $\frac{57.3}{a} + \frac{3a}{1000} : 1 =$ hypotenuse : opposite side. For since the equimultiples of quantities bear the same ratio, to each other as the quantities themselves, if the two first terms be multiplied by a, the equality, will stand thus: $57.3 + \frac{3aa}{1000} : a =$ hypotenuse : opposite side.

These are the chief rules hitherto given for the 4 cases of rectangular trigonometry.

Oblique triangles are treated after the same manner, by first dividing them into 2 right-angled triangles, by means of a perpendicular, which must always fall from the end of a given side, and opposite to a given angle. Here, let it be observed: (1) The perpendicular will sometimes fall within, and sometimes without the triangle; when it falls within, it falls upon some part of the base, or longest side, but when it falls without, it falls upon one of the shorter sides continued. In either case, there are two right-angled triangles made, whose sides or angles are to be found as above. (2) The segments of the base, occasioned by the perpendicular falling upon it, are easily calculated by the following proportion. As the base is to the sum of the other two sides, so is the difference of the sides to the difference of the segments of the base. The half of this difference added to half the base gives the greater segment, and subtracted from it, the less,er,.

Introduction to Spherical Geometry

Horatio Nelson Robinson, 1854

Some older writers (like Edward Wells, of the last extract but one) left spherical geometry aside as less useful or too much more difficult than its plane counterpart, but it had been clear since at least the sixteenth century that accurate navigation over long distances, as well as a whole range of astronomical calculations, must necessarily depend on it.

Horatio Nelson Robinson was the author of a string of textbooks published in the United States in the 1850s and 1860s on the various branches of mathematics: arithmetic, algebra, astronomy, and also natural philosophy and the practical disciplines of surveying and navigation. Some were aimed at beginners, others at university students, but all took a briskly practical approach, sweeping away "rubbish," "redundancies" and, particularly in geometry, "attempts to prove what is perfectly obvious," as Robinson put it.

Little is known of Robinson himself; his name points to a family connection with the Royal Navy, but without knowing *when* he was named after the Great British hero it is difficult to say more.

Horatio Nelson Robinson, *Elements of Geometry, Plane and Spherical Trigonometry, and Conic Sections* (Cincinnnati: 16th edition, 1854), pp. 176–178.

Spherical Geometry is nothing more than the general principles of geometry applied to the various sections of a sphere; and spherical trigonometry is but the general principles of plane trigonometry applied to triangles resting on a surface of a sphere, and the planes of the sides of the triangles passing through the center of the sphere.

Definitions

1. A sphere is a solid whose surface is equally convex in every part, and every point of the surface is equally distant from one point within, and this point is called the center. A sphere may be conceived to be generated by the revolution of a semicircle about its diameter.

If the center of the semicircle rests at the same point, the position of the diameter may be in any direction or position, and the revolution of the semicircle will describe the same sphere.

2. Any plane that passes through the center of the sphere divides the solid and the surface into two equal parts.

3. Any two planes that pass through the center of a sphere intersect each other on the opposite points of the sphere, because the intersection of any two planes is a right line.

4. A *great circle* on a sphere is one whose plane passes through the center of the sphere.

5. Every great circle has *poles*: two points on the sphere directly opposite to each other and equally distant from every point on the great circle.

The distance from any pole to its equator, in any direction, is one fourth of the whole distance round the sphere.

6. Any point on a sphere may be a pole to some great circle.

7. A spherical triangle is formed by the intersection of three great circles on a sphere. Conceive three radii drawn from the three angular points to the center of the sphere, thence forming a solid angle. The angles of the three planes which form this solid angle at the center are the three angles which measure the sides of the triangle, and the inclination of these planes to each other form the angles of the triangle.

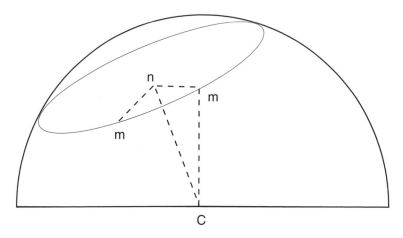

Figure 4.9. Every intersection of a sphere by a plane is a circle.

8. The complete measure of a spherical triangle is but the complete measure of a solid angle at the center of a sphere; and this solid angle is the same, whatever be the radius of the sphere.

9. Every great circle, or portion of a great circle on the surface of a sphere, has its poles; conversely, every pole, or the point where two circles intersect, has its equator 90° distant, and the portion of this equator between the two sides, or the two sides produced, measures the spherical angle at the pole.

The inclination of two tangents of two arcs, formed at their point of intersection, also measures the spherical angle.

10. We can always draw one and only one great circle through any two points on the surface of a sphere; for the two given points and the center of the sphere give three points, and through three points only one plane can be made to pass.

Proposition 1. Every ⌞inter⌟section of a sphere by a plane is a circle

If the plane passes through the center of the sphere, the ⌞inter⌟section is evidently a circle, for every point on the surface of the sphere is equally distant from the center. These ⌞inter⌟sections are great circles, and all great circles on the same sphere are equal to each other.

Now let the cutting plane not pass through the center. From the center C, let fall Cn perpendicular to the plane (see Figure 4.9); and when

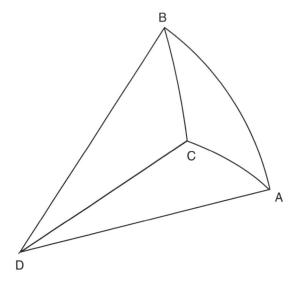

Figure 4.10. Any two sides of a spherical triangle are together greater than the third.

a line is perpendicular to a plane, it is perpendicular to all lines that can be drawn in that plane. Therefore, any line, as nm, in the plane is at right angles to Cn. Hence $nm = \sqrt{Cm^2 - Cn^2}$.

But nm is *any* line in the plane, from the point n to the surface of the sphere. So, because, this value for nm is invariable ⌞...⌟ it is the radius of a circle whose center is n.

N.B. These circles are called small circles, and are greater or less, as they are nearer or more remote from the center C.

Small circles on a sphere are never considered as sides of spherical triangles. We again repeat, that sides of spherical triangles must be portions of *great* circles, and each side must be less than $180°$.

Proposition 2. Any two sides of a spherical triangle are together greater than the third

Let AB, AC, and BC be the three sides of the triangle, and D the center of the sphere (see Figure 4.10).

The arcs AB, AC, and BC are measured by the angles of the planes that form the solid angle at D. But any two of these angles are together greater

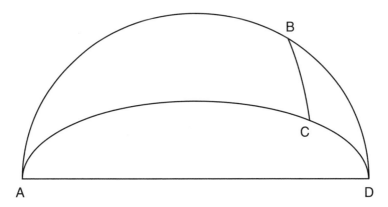

Figure 4.11. The sum of the three sides of any spherical triangle is less than the circumference of a great circle.

than the third. Therefore, any two sides of the triangle are, together, greater than the third. *Q.E.D.*

Proposition 3. The sum of the three sides of any spherical triangle is less than the circumference of a great circle

Let *ABC* be a triangle (see Figure 4.11); the two sides *AB, AC*, produced, will meet at the point on the sphere which is directly opposite to *A*, and the arcs *ABD* and *ACD* are together equal to a great circle. But by the last proposition, *BC* is less than the two arcs *BD* and *DC*. Therefore, *AB, BC* and *AC* are together less than *ABD + ACD*, that is, less than a great circle. *Q.E.D.*

Napier's Rules

Alan Clive Gardner, 1956

Alan Clive Gardner had presumably been, at one stage, one of the second mates at whom the present book was aimed, most likely in the British merchant navy. He was also the author of a work on navigation. The book—which in a revised form is still in print—was written to introduce those mathematical principles

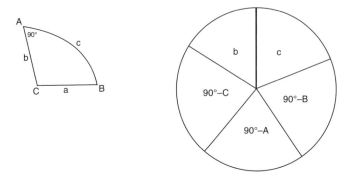

Figure 4.12. Napier's Five Circular Parts.

which Gardner felt should be familiar to "a modern ship's officer." This section on Napier's rules (we will meet Napier, or at least his bones, again in Chapter 6) begins to point toward the practical use of some of the geometry we have met in this chapter.

A. C. Gardner (dates unknown), *Mathematical Notes and Examples for Second Mates. Part II: Trigonometry, Mensuration and Graphs.* (Glasgow, 1956), pp. 60–63. Reproduced by the publishers, Brown, Son & Ferguson, Ltd., 4–10 Darnley St., Glasgow, G41 25D, Scotland, with their kind permission.

Napier's Rules. The Solution of Right-Angled and Quadrantal Spherical Triangles

A right-angled spherical triangle is a spherical triangle in which one of the angles is 90°.

If we are given any two parts of the triangle (in addition to the right angle), we can find the other parts. By "parts" are meant sides and/or angles.

Method of Solution

Draw the triangle, name the angles and sides, and then prepare "Napier's Five Circular Parts." This is done as follows.

Draw a circle and divide it into five sectors (see Figure 4.12). Starting from the right angle, which is indicated by a heavy line, give each sector the name of one part of the triangle, naming the parts in order round the circle. The two parts next to the right angle are named normally, but the other three parts are replaced by their complements.

Two Rules must be memorized in connection with the Circular Parts. They are as follows:

1. Sine Middle Part = Product of Cosines of Opposite Parts.
2. Sine Middle Part = Product of Tangents of Adjacent Parts.

Examples on the Use of These Rules

(1) Given C and a, to find c.

Here the three parts we are dealing with are connected by Rule I above, so we write:

$$\sin c = \cos(90° - C).\cos(90° - a),$$

which becomes:

$$\sin c = \sin C.\sin a.$$

(2) Given C and B, to find a.

Here the three parts are connected by the second rule, so we write:

$$\sin(90° - a) = \tan(90° - C).\tan(90° - B),$$

which becomes:

$$\cos a = \cot C.\cot B.$$

The remaining examples are harder.

(3) Given a and c, to find b.

Here the three parts are connected by the first rule, because $(90° - A)$ is the "middle part" which is opposite to b and c. So we write:

$$\sin(90° - A) = \cos b.\cos c.$$

Remove complement:

$$\cos a = \cos b.\cos c.$$

Now transpose formula so as to bring $\cos b$ on to L.H.S.

$$\cos b = \frac{\cos a}{\cos c}$$

$$\cos b = \cos a.\sec c$$

The last step is desirable because it is always better to add logs than to subtract them .... There is less chance of error.

(4) Given C and a, to find B.

These three parts are connected by the second rule, so we write:

$$\sin(90 - a) = \tan(90 - C).\tan(90 - B)$$

whence: $\cos a = \cot C.\cot B$

$$\therefore \cot B = \cos a.\tan C$$

The Rule of Signs

In the numerical examples which follow, the "Rule of Signs" must be observed. This rule will be explained as it is applied in each example.

Example

A is the North Pole, C is a point on the Greenwich Meridian in the Northern Hemisphere, and B is on the Meridian of 90° E. The initial Great Circle track from C to B is 102°, and the inital Great Circle track from B to C is 290°. What is the Great Circle Distance from B to C ...?

To find a, the Great Circle Distance from C to B

$$\sin(90° - a) = \tan(90° - C).\tan(90° - B)$$

$$\therefore \cos a = \cot C.\cot B$$

$$\therefore \cos a = \cot\overset{-}{102°}.\cot\overset{+}{70°}$$

By the Rule of Signs, minus multiplied by plus equals minus, ∴ cos a is negative, so "a" must be in the 2nd Quadrant.

Therefore, whatever angle we get from the Tables for "a" must be subtracted from 180° to give the correct value.

From the Tables, "a"=85°34'. But cos a is negative.

$$\therefore a = 180° - 85°34' = 94°26'$$

∴ Great Circle Distance from B to C = 5666 nautical miles.

◆ 5 ◆

Maps, Monsters, and Riddles: The Worlds of Mathematical Popularization

POPULARIZING MATHEMATICS IS A GAME WITH TWO HALVES. THE FIRST HALF took place in the early eighteenth century, in a world amazed by the phenomenal success of Newtonian science and, perhaps, anxious about what mathematics now meant and did. Quite a lot of the popular mathematical writing of this period appeared in the specific context of explaining Newtonian science to the nonspecialist, and our extract from the lovely *Newton for the Ladies* illustrates both this and what would much later become a vitally important line in popular science writing: "it can be proved that ...." Trust me, I'm a mathematician.

Real popularization of mathematics, different from games and puzzles but also different from textbooks or self-improvement manuals is not very prominent (in English) in the later eighteenth and nineteenth centuries; my guess is that an ethic of virtuous, self-improving leisure might have acted to produce nonspecialist mathematics writing of a rather different character, like *The Juvenile Encyclopedia* (Chapter 6) or *The Popular Educator* (Chapter 2). So we take up our story again around the beginning of the twentieth century. This period saw an explosion of popular writing, and, as our extracts show, it ranged across the whole gamut of mathematics, from the most pure—number theory, say—to the most applied—Newton's and Einstein's laws of motion.

This wave of writing responded in part to the sorts of concerns we will meet in Chapter 11: that mathematics and science had dehumanized the world or placed too much power in the hands of a few specialists. It also responded to the new phenomenon of "mathematics anxiety." Mathematics had always been hard, and harder for some than for others, but the twentieth century saw the emergence of the self-conscious feeling that the subject had an image problem, and many of the popularizers of the twentieth century saw themselves as specifically acting to correct the situation. As Dan Pedoe put it, "not all mathematicians are witty and clever."

The Athenian Mercury

1691–1697

Something of a publishing sensation in its day, *The Athenian Mercury* was a kind of seventeenth-century quiz show, a weekly newspaper which ran from 1691 to 1697, devoted to answering queries sent in by readers. The answers were supplied by a small group of writers, of whom one, Richard Sault (d. 1702), was a mathematician and mathematics teacher. The quantity of mathematical material in the *Mercury* was never very great, but it provides a unique window on the concerns of amateur mathematicians in the Britain of the 1690s. The weekly issues were republished in volume form, under the title *The Athenian Oracle*, in the first decade of the eighteenth century; this is the source for these extracts.

The Athenian Oracle: being an entire collection of all the valuable questions and answers in the old Athenian mercuries. ⌐. . .⌐ *By a member of the Athenian Society.* (London, 1703–1704 (vols. I and II); 3rd edition, 1716 (vol. III)); vol. I, pp. 149, 355–356; vol. II, pp. 242, 306–307; vol. III, pp. 248–249, 398–400.

> *Q.* Gentlemen, I am resolv'd to Get round the Earth on Foot; I desire to know whether my Head or Feet will travel most, and how much one more than the other?
>
> *A.* Pray Sir, which way do you design to travel, that you'll meet with no Water, Mountains of Ice, etc? However, Sail or Go, you are desired to tell us how large a Circle you design to take; as, also, as near as you can, your height; but besides all this, (which is yet a greater task) pray send to us the way of squaring a Circle. If you cannot do that, we assure you we can't answer the Question exactly, and for mathematicians to advance anything that won't bear a demonstration, is worse than doing nothing at all.
>
> *Q.* Whether One be ⌐a⌐ Number?
>
> *A.* Diophantus,° that Prince of Arithmeticians, calls it a Number, and we take it to be so too. Some says 'tis rather the *Genesis*, or beginning of Numbers, than a Number itself, since all other Numbers are made

out of it; but that is to make it both Integer and Fraction at once, which is impossible.

Q. Having lately read in one of the Books of Dee's Euclid° something concerning perfect Numbers, and it not being in my Capacity (being but a young Student) to comprehend the true Notion thereof, I beg your Assistance in it, so far as to satisfy what they are, and by what Means I may find any one of them out, for I find them to be of great Use to me, and in so doing you'll highly oblige your Friend, etc.

A. A perfect Number is that which is equal to all its aliquot Parts added together. According to this Definition, 6 is a perfect Number, because if you take its aliquot Parts, which are 1, 2, 3, their Sum will be equal to 6; again, 28 is a perfect Number, because its aliquot Parts, 1, 2, 4, 7, 14, added together make 28. Now, if you will find as many of 'em as you please, take the following Progression: 1, 2, 4, 8, 16, 32, etc., which it is easy to continue in doubling every last Term. Choose in this Progression any one Term; subtract Unity from it; if the Remainder is a prime Number, multiply this Remainder by the Term immediately preceding. The Product will be a perfect Number. But if the Remainder is no prime Number you must choose another Term. This Rule will be cleared by some Instances: take the Term 4; subtract Unity from it; the Remainder is 3, which being multiplied by the Term immediately preceding, *viz.* 2, the Product, 6, is a perfect Number. Again, take the Term 8; subtract Unity from it; the Remainder is 7. Multiply this Remainder by 4; the Product is 28, which is a perfect Number. But if you would take 16, because, having taken Unity from 16, the Remainder, 15, is no prime Number, the Product of 15 by 8 won't be a perfect Number. Therefore take the following Term, 32, and, working as is prescribed, you will find 496 for another perfect Number.

This is a very easy Rule, and we expect our Querist will be pleased with it; but we have something of far greater Consequence to him, which is that we, guessing by his Question that he is apt to attribute some Virtue to perfect Numbers (or else why should he think they can be of great Use to him?), and ‚the‚ Doctrine of attributing Virtue to Numbers being a pure *Chimera* of *Cabalistic* Sprits, we advise him to employ his Time better, than in such a vain fruitless Contemplation.

Q. *1.* If in an Arithmetic Progression from Unity, the last Number be thirty-six, and the Sum of all the Numbers be one hundred and forty eight, how many Terms are there in all?

Q. *2.* And if eight Numbers be in Arithmetical Proportion from Unity, the last Term thirty-six, what is their common Difference? And how may these two be discovered and resolved?

A. In the first Question the Number of Terms is Eight, and in the second the Difference is Five. Both resolved by this Method.

Let *a* be equal to the first Term, *t* equal to the last Term, *n* equal to the Number of Terms, *d* equal to the difference, and *s* equal to the Sum of all the Terms. Then by this Theorem, $S = \frac{1}{2}as + \frac{1}{2}nt$, it will be found $p = \frac{2s}{a \times t}$. And, *s*, *a*, and *t* being already given, *n* will be discovered to be equal to eight, the number of Terms sought. And in the second Question, where the Difference of Terms is required, *t*, *a*, *n* being given, by this Theorem—$t - a + dn = d$—it will be $d = \frac{t-1}{n-1} = 5$, equal to the difference.

Q. Reading the other Day in Wallis's Algebra,° I found at Page 65 this Question, *viz.*, To find three Numbers, whereof the first with $\frac{1}{3}$ of the other two shall make 14, and ˌtheˌ second with $\frac{3}{4}$ of the otherˌs shall makeˌ 8, and the Third with $\frac{2}{3}$ of the otherˌs shall makeˌ 8.

I understood the Question with all its Reasons till I came a little farther, where he says,

Then multiply the second and third by 3, and making subductions to destroy 2, or the first (in the like manner as Dr. Pell in his Algebra directs to do).

And here I stopped, not being able to comprehend the Reasons of the last mentioned multiplying, etc., nor what followed. Therefore I desire the Question solved with all its Reasonˌsˌ, may be in the next Oracle, which will oblige, yours, J.M.

A. We can't have ˌtimeˌ to look in Dr. Wallis's Algebra; the Question may be briefly solved thus:

Let the 3 Nˌumbersˌ be *x*, *y*, *z*; then according to the Tenor of the Question,

$$x + \frac{y + z}{3} = 14, \tag{1}$$

$$y + \frac{x+z}{4} = 8, \qquad (2)$$

$$z + \frac{x+y}{5} = 8. \qquad (3)$$

Which being reduced out of your Fraction will stand thus,

$$3x + y + z = 42, \qquad (1)$$

$$4y + x + z = 32, \qquad (2)$$

$$5z + x + y = 40. \qquad (3)$$

Now it remains to destroy all the unknown Terms but one. Subtract the second Equation out of the first to destroy z: then $2x - 3y = 10$. Next multiply the second Equation by 5, because of making z in each equal to $5z$, *viz.* $20y + 5x + 5z = 160$. Out of which subtract the 3rd Equation *viz.*:

$$20y + 5x + 5z = 160, \qquad (2)$$

$$5z + x + y = 40. \qquad (3)$$

There remains (1) $19y + 4x = 120$, and we had before (2) $2x - 3y = 10$.

So that now we have two Equations, wherein z is destroyed. Multiply the last by 2, that there may be $4x$ in both, and then subtract to destroy x also, *viz.*,

$$19y + 4x = 120, \qquad (1)$$

$$4x - 6y = 20. \qquad (2)$$

Whence $y = \frac{100}{25} = 4$, having found $25y = 100$.

Of the three, the other two are also discovered from the simplest of the preceding Equations, *viz.*, $x = 11$ and $z = 5$.

We suppose the Reason of the Doctor's "Multiplying the second, and the third, by 3, to destroy a," as you say, was to make the Products of a Equal, that thereby it might be made capable of being destroyed by Subtraction.

Notes

Diophantus of Alexandria (3rd century AD): Greek-speaking mathematician best known for his *Arithmetica*, a collection of problems equivalent (mostly) to linear, quadratic, and simultaneous equations.

Dee's Euclid: Henry Billingsley's 1570 translation of Euclid, the first into English. See Chapter 7 for an extract from the translation and Chapter 10 for a passage from John Dee's famous preface.

Wallis's Algebra: the *Treatise on Algebra* (1685), by the English mathematician John Wallis (1616–1703).

Newton for the Ladies

Francesco Algarotti, 1739

Cast in the form of a dialogue between a "Marchioness" and the narrator, most of this book is concerned with Newton's ideas about optics, but toward the end there is some material about universal gravitation. Here we see the narrator (almost as overbearing as Robert Recorde's "Master") explain Kepler's second law and begin to discuss a principle of universal attraction; between them these two mathematical laws of nature framed the narrative of Newtonian discovery as far as it concerned celestial dynamics.

Algarotti was an Italian Newtonian, a friend of Voltaire, and a fellow of the Royal Society of London. This book, part of a wave of Newtonian popularization in the first half of the eighteenth century, was translated shortly after its first appearance by Elizabeth Carter (1717–1806), a poet and writer, and later a Greek scholar in her own right.

Francesco Algarotti (1712–1764), *Sir Isaac Newton's philosophy explain'd for the use of the ladies. In six dialogues on light and colours. From the Italian of Sig. Algarotti.* (London, 1739), vol. 2, pp. 151–153, 159–163.

The five Planets in whose Number we may safely replace our Earth, are called *Primary*, to distinguish them from other subaltern Planets which revolve round a Primary as the Moon does round our Earth, the four Satellites of *Jupiter* round that Planet, and five round *Saturn*, and these Subalterns are called *Secondaries*. These last agree with their Primaries in observing this Order, that the nearest complete their Orbit in a less, and the most distant in a greater Time; and they keep this Law° with the same Exactness and the same Relation as their Primaries do.

Another Law in which the primary and secondary Planets agree is, that they do not describe equal Parts of Orbits in equal Times, but such Parts of Orbits as to make their Areas equal. That you may better understand this other Law of their Motion, you are to suppose that the Orbit of a primary Planet is very near a Circle which the Sun is not placed directly in the Center of, but a little on one Side. Imagine a Line to be drawn from that Point of the Orbit where the Planet now is, to the Sun, and another Line to be drawn from that Point where the Planet will be twenty-four Hours hence. The Space contained betwixt the two Lines drawn to the Sun, and that Part of the Orbit which the Planet has described in twenty-four Hours, is called the *Area*, and will be equal to another such Area, that the Planet will describe in twenty-four Hours more; and thus in equal Times the Areas will be always equal.

The *Areas* then, as Astronomers express it, are *proportional to the Times*. Thus if, instead of twenty-four Hours, we put twelve, which is one Half, an Area described in those twelve Hours will be only one Half of the Area described in the 24; and so, if we take a third or fourth part of the Time, the Areas described in that third or fourth Part will be the third or fourth of those described in the first Time; and if that Time be doubled, the Area described in it will be doubled likewise, and so on. This Law which the primary Planets observe with Regard to the Sun, the secondary Planets° observe with Regard to the Primary, round which they revolve; and this Primary is the same to its Satellites as the Sun is to the Planets of the first Order.

ᴏ̴

Sir *Isaac Newton*, continued I, founded his Scheme in Geometry, which we may call his native Country. He began with demonstrating that if a Body in Motion is attracted towards a Point either moveable or immoveable, it will describe about this Point equal Areas in equal Times, and in general, that the Areas will be proportional to the Times; and on the contrary, if a Body describes round a moveable or immoveable Point Areas proportional to the Times, it will be attracted towards that Point. That is, the Body will have such a Tendency towards the Point, that if every other Motion which impels it a different Way should cease, the Body would directly unite itself to that Point, just as Bodies here below, when left to themselves, fall directly upon the Earth.

This Principle, interrupted the Marchioness, is equally applicable both to the primary and secondary Planets. Each of these describe Areas proportional to the Times, round the Point about which they turn (if the Sun, our Earth and Jupiter may be termed Points). The primary Planets then are attracted by the Sun, and the secondary by their respective Primaries about which they revolve. Is not this a necessary Consequence?

It is without Dispute absolutely necessary, answered I. But remember, Madam, this is a Deduction of your own. ⸢....⸣

You say then that there is a Force in the Sun which attracts the Planets to him, and after the same Manner, a Force in the Planets that attracts the Satellites, and this attractive Force, joined with another by which they all move from West to East, is the Reason why they first revolve round the Sun, and the others round their Primaries, in a certain Order.

The Ancients, in order to explain this difficult Phenomenon, built solid Heavens and created Intelligences to put them in Motion; on the other Hand, Descartes had embarrassed the whole Universe with the great and magnificent Apparatus of his Vortices.° But after all, the Motion of the heavenly Bodies is by Sir Isaac Newton reduced to the most simple, yet the most noble Phenomenon in the World, which has been rendered much more familiar in Europe than is agreeable to some Persons. In short, it is no more than that of a Bullet, which would of itself proceed in a direct Line, if the attractive Force of the Earth did not oblige it to move in a Curve. The Bullet very soon falls to the Earth, because the greatest Force we can possibly give it is but little when compared to the vast Extent of this Globe. If it were possible for human Weakness to throw one from hence beyond Peru, it is demonstrated that we should acquire a new Satellite; it would like the Moon revolve about our Earth, only its Motion being necessarily very soon weakened from the continual Resistance of the Air, while the gravitating Force would lose nothing of its Strength, this new Moon would at last fall and destroy every thing it lighted on, after we had heard it make a horrible hissing over our Heads.

Notes

this Law: Algarotti refers to Kepler's third law, which he has previously explained: the square of a planet's orbital period is proportional to the cube of the orbit's longest diameter.
secondary Planets: the satellites (moons) of the various planets.

Vortices: René Descartes (1596–1650) had attempted to explain the motions of the planets
on the hypothesis that apparently empty space was in fact filled with an "ether," whose
vortices could account for the observed orbital behavior of bodies.

Maps and Mazes

W. W. Rouse Ball, 1892

First published in 1892, Rouse Ball's *Recreations* has become one of the classics
of popular mathematics. In the preface Rouse Ball—he was a well-known fellow
and lecturer at Trinity College, Cambridge—remarks: "I hasten to add that the
conclusions are of no practical use, and most of the results are not new," features
which the book has in common with the works of predecessors like van Etten
and Leybourne. But in other ways it is very different from their books of games
and amusements, and points toward the mathematics popularization genre of
the twentieth and twenty-first centuries. Rouse Ball gives detailed references to
historical treatments of the problems and devotes a substantial portion of the book
to sophisticated accounts of the ancient problems of duplicating the cube, trisecting
the angle, and squaring the circle, often—as in the following extracts—explaining
multiple approaches or proofs for the same problem. The book as a whole thus also
has something of the character of a history of recreational mathematics.

W. W. Rouse Ball (1850–1925), *Mathematical Recreations and Problems of Past and Present
Times* (London and New York: 2nd edition, 1892), pp. 36–38, 129–131, 159–162.

Colouring Maps

I concluded the last chapter by stating one or two arithmetical theorems
of which hitherto no proofs have been given. As a pendant to them I may
mention the geometrical proposition that *not more than four colours are
necessary in order to colour a map of a country (divided into districts) in
such a way that no two contiguous districts shall be of the same colour.* By
contiguous districts are meant districts having a common *line* as part of
their boundaries; districts which touch only at points are not contiguous in
this sense.

The earliest enunciation of this problem seems to be due to Francis
Guthrie, who communicated it to De Morgan about 1850, but it is said

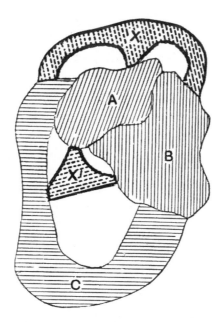

Figure 5.1. Three contiguous districts, and another district contiguous with all of them. (Rouse Ball, p. 37. ©The Bodleian Libraries, University of Oxford. Jessel e.91.)

that the fact had been familiar to practical map-makers for a long time previously. It was through De Morgan that the proposition became known generally, and in 1878 Cayley° recalled attention to the proposition by stating that he did not know of any rigorous proof of it.

Probably the following argument, though not a formal demonstration, will satisfy the reader that the result is true.

Let A, B, C be three contiguous districts, and let X be any other district contiguous with all of them. Then X must lie either wholly outside the external boundary of the area ABC or wholly inside the internal boundary, that is, it must occupy a position either like X or like X' (see Figure 5.1).

In either case every remaining unoccupied area in the figure is enclosed by the boundaries of not more than three districts: hence there is no possible way of drawing another area Y which shall be contiguous with A, B, C, and X. In other words, it is possible to draw on a plane four areas which are contiguous, but it is not possible to draw five such areas.

If A, B, C are not contiguous, each with the other, or if X is not contiguous with A, B, and C, it is not necessary to colour them all differently, and thus the most unfavourable case is that already treated. Moreover any of the above areas may diminish to a point and finally disappear without affecting the argument.

That we may require at least four colours is obvious from the above diagram, since in that case the areas A, B, C, and X would have to be coloured differently.

Prof. Tait° asserts that, since three colours are sufficient to colour a set of lines joining an even number of points and such that three and only three lines meet at each point so that no two conterminous lines are of the same colour, it follows that four colours will suffice for a map.

The following argument, by which the truth of the proposition has been supposed hitherto to be demonstrated strictly, was given by Mr A.B. Kempe.° It rests on proving (i) that every map must have at least one pair of adjacent districts having only one common boundary; (ii) that every map must have one district with less than six boundaries; (iii) that it will be sufficient to prove the truth of the proposition for a map reduced by throwing two of these bounding districts into one; (iv) that if the colours are permuted properly one more reduction similar to (iii) may be made. If these are granted the proposition follows, but his demonstration of the fourth step in the argument is not valid.

In his original paper Mr Kempe added the two additional theorems: (i) that if not more than three boundaries meet at any point and if (excluding islands and peninsulas) every district touches only an even number of other districts, then three colours will suffice; and (ii) that if an even number of boundaries meet at every point, then two colours will suffice.

Mazes

Every one has read of the labyrinth of Minos in Crete and of Rosamund's bower. A few modern mazes exist here and there—notably one, which is a very poor specimen of its kind, at Hampton Court—and in one of these, or at any rate on a drawing of one, most of us have threaded our way to the interior. I proceed now to consider the manner in which any such construction may be traversed completely even by one who is ignorant of its plan.

The theory of the description of mazes is included in Euler's theorem on unicursal curves.° The paths in the maze are what previously we have termed branches, and the places where two or more paths meet are nodes. The entrance to the maze, the end of a blind alley, and the centre of the maze are free ends and therefore odd nodes.

If the only odd nodes are the entrance to the maze and the centre of it—which will necessitate the absence of all blind alleys—the maze can be described unicursally. ⌊...⌋ Again, no matter how many odd nodes there may be in a maze, we always can find a route which will take us from the entrance to the centre without retracing our steps, though such a route will take us through only a part of the maze. But in neither of the cases mentioned in this paragraph can the route be determined without a plan of the maze.

A plan is not necessary, however, if we make use of Euler's suggestion and suppose that every path in the maze is duplicated. In this case we can give definite rules for the complete description of the whole of any maze even if we are entirely ignorant of its plan. Of course to walk twice over every path in a labyrinth is not the shortest way of arriving at the centre, but, if it is performed correctly, the whole maze is traversed, the arrival at the centre at some point in the course of the route is certain, and it is impossible to lose one's way.

I need hardly repeat that the complete description of such a duplicated maze is possible, for now every node is even, and hence, by Euler's second proposition, if we begin at the entrance we can traverse the whole maze, in so doing we shall at some point arrive at the centre, and finally shall emerge at the point from which we started. This description will require us to go over every path in the maze twice, and as a matter of fact the two passages along any path will be made always in opposite directions.

In order to describe such a maze without knowing its plan it is necesary to have some means of marking the paths which we traverse and the direction in which we have traversed them—for example, by drawing an arrow at the entrance and end of every path traversed, or better perhaps by marking the wall on the right-hand side, in which case a path may not be entered when there is a mark on each side of it.

Of the various practical rules for threading a maze those enunciated by M. Trémaux° seem to be the simplest. These I proceed to explain. For brevity I shall describe a path or a node as old or new according as it has been traversed once before or not at all. Then the rules are (i) whenever you come to a new node, take any path you like; (ii) whenever you come

by a new path to an old node or to the closed end of a blind alley, turn
back along the path by which you came; (iii) whenever you come by an old
path to an old node, take a new path, if there is one, but if not, an old path;
(iv) of course a path traversed twice must not be entered. I should add that
on emerging at any node then, of the various routes which are permitted
by these rules, it will be convenient always to select that which lies next to
one's right hand, or always that which lies next to one's left hand.

Notes

Cayley: Arthur Cayley (1821–1895), probably best known for his work on matrices.

Tait: Peter Guthrie Tait (1831–1901), Scottish mathematician and physicist.

Kempe: Alfred Bray Kempe (1849–1922), still best known for this (false) proof, on which the
1976 computer-aided proof of the four-color theorem was based.

Euler's theorem: What Rouse Ball calls Euler's theorem evidently says that on a map
comprising paths ("branches") and their meeting points ("nodes"), it is possible to walk
every path without ever treading the same path twice ("unicursally") only if the number
of nodes at which odd numbers of paths meet ("odd nodes") is two or fewer. If it is zero,
then it is possible to walk every path without repetition *and* end where you started; this
is Euler's second proposition, mentioned a little later.

M. Trémaux seems now to be entirely unknown apart from these rules for threading a maze.

"Einstein's Real Achievement"

Oliver Lodge, 1921

Oliver Lodge was a physicist and a prominent member of the British Association;
here he writes on one of the important scientific subjects of the day, the theory of
relativity, successor to Newton's account of dynamics and gravitation. Ironically,
Newton himself had argued against the space-filling "ether" to which Lodge
appeals to explain relativistic effects.

Ultimately the ancestors of such scientific magazine articles lay in the mathematical and scientific sections of almanacs and periodicals like *The Ladies' Diary*
as much as in popular books like *Newton for the Ladies*.

Oliver Lodge (1851–1940), "Einstein's Real Achievement," *The Fortnightly Review*, 1 September, 1921, pp. 360–362.

The Lorentz transformation is of fundamental importance, and it is
worth an effort to understand it. The equations expressing it have been
arrived at in a great number of ways, which indeed is a test of their truth.

Let us see if we can partially explain, in rather more detail, how those equations may be arrived at. If a medium is in motion relatively to bodies in it, and we want to refer our motion to that medium instead of to a relatively fixed standard body, we have to modify our expression for distance. Think of an engine standing still at a distance a from a railway station, and let a man be lying on the line at a greater distance x from the same station. Referred to the station his position is defined by x, but referred to the engine his position is $x - a$. Now let the engine begin moving towards him with velocity u, so that a is no longer constant, but increases with the time, $a = ut$; it becomes imperative that the man estimates his distance from the engine instead of from the station, and his static distance $x - a$ has now become the kinematic $x - ut$.

That is called changing the origin or standard of reference, and is typical of the simplest part of what has to be done in physics when we try to attend to our relative motion with respect to the ether of space. The frame of reference is in relative motion with reference to us, or, more precisely, we are in relative motion with respect to it—which for calculation purposes is the same thing.

Furthermore we find that we have theoretically, and practically when it comes to great distances and great speeds, to consider by what means we become aware of the position of the engine or other moving body. If we estimate it by sound, we willingly allow for the time taken up by the messenger on the journey, and we know that the real distance at each instant is less than it appears to be by hearing. If we depend for our information on our sense of sight, the messenger is very rapid, and we are usually content with estimating an approaching distant object as where it *appears* to be, which, strictly speaking, is rather further off than it really is. So also we perceive an occurrence on it as if it happened rather later than it really does. For instance, the reading of a distant clock-face is necessarily a trifle belated, and our perception of the bursting out of a new star may be a century behind. In that interval of time the star may have approached us a great deal nearer than our measurements give.

So we can admit that in order to record times and places in a way which is independent of our own position and relative movement, and which will be intelligible to people anywhere, and so to speak "true" at the moment we record them, we must allow for the time taken by light to reach us. Now that time will be $\frac{x}{c}$ if the thing observed is relatively stationary, and $\frac{x'}{c}$ or

$\frac{x-ut}{c}$ if it be relatively approaching, c being the velocity of light. So calling this corrected time t', corresponding to the corrected distance x',

$$t' = \frac{x - ut}{c} = \frac{x}{c} - \frac{u}{c}t = t - \frac{ux}{c^2},$$

which after all is only the fairly obvious $\frac{t'}{t} = \frac{c-u}{c} = \frac{x'}{x}$. This gives us the true values, x' and t', for place and time of an approaching object which to the observer appears to be at x and t.

But now comes a curious and unexpected complication, such as cannot be illustrated by railways and common experience, and such as could not be anticipated by mere kinematics. All that has been said so far is true of abstract motion, but when we come to physical motion through an ether, there is another unexpected effect to be taken into account—at least whenever a body composed of a group of electrical particles is moving. A group of electric charges cannot move together ⌐...⌐ without to some extent affecting their positions relative to each other. They tend to crowd together along the line of march, and perhaps spread out a little sideways. ⌐...⌐

The expressions that would otherwise be correct have, therefore, to be multiplied by a factor β, which is very nearly unity save for excessive speeds approaching the medium's critical value c. At such speeds as that, the medium's properties are become strained or exhausted. It cannot transmit anything with a speed greater than c, and the coefficient β rapidly approaches an infinite value as the speed c is approximated to. For all ordinary speeds, however, it is very nearly 1. We thus arrive at the equations recorded above:

$$x' = \beta(x - ut);$$

$$t' = \beta(t - \frac{ux}{c^2});$$

$$\text{with } \beta^2(c^2 - u^2) = c^2.$$

and now no one can tell whether it be the source or the observer that is moving.

The gist of the equations is that a moving observer must take not only his distances as variable, but his times too. He must have a local and fictitious time if he is to ignore his own motion and treat his direct measurements as conclusive.

Einstein's step was to dispense with any fiction about this subjective or local measure of time, to claim that it was as real as any other, and to see what happened.

Riddles in Mathematics

Eugene P. Northrop, 1945

Eugene Northrop is now a somewhat obscure writer; he studied at Yale in the 1930s and seems to have published no other books. His volume of *Riddles in Mathematics* takes an idiosyncratic look at the lighter side of mathematics, written for, as he said, those interested in the subject as a recreation from "the human equation" of everyday life. But the discussion of paradoxes about probability shown here takes a philosophically serious approach to that subject which contrasts with the naive reckoning of dice and coin-tossing games found in some earlier popular explorations of the subject. More light will be shed on the paradoxical infinity of the real numbers by Dan Pedoe later in this chapter.

Eugene P. Northrop (b. 1908), *Riddles in Mathematics: A Book of Paradoxes* (London, 1945), pp. 74, 166–168.

Algebraical fallacies

Paradox 1.

1 cat has 4 legs; (1)

no (i.e., 0) cat has 3 legs. (2)

Adding the "equals" (1) and the "equals" (2), we conclude that 1 cat has 7 legs.

Paradox 2.

$$2 \text{ pounds} = 32 \text{ ounces;} \tag{1}$$

$$\frac{1}{2} \text{ pound} = 8 \text{ ounces.} \tag{2}$$

Multiplying the equals (1) by the equals (2), we obtain 1 pound = 256 ounces.

Paradox 3.

$$1 \cdot 0 = 2 \cdot 0; \tag{1}$$

$$0 = 0. \tag{2}$$

Dividing the equals (1) by the equals (2) gives $1 = 2$.

Paradox 4.

$$(-a)^2 = (+a)^2,$$

since the square of a negative quantity is positive. Extracting the square root of both sides, we have $-a = +a$.

Paradox 5.

$$\frac{1}{4}\text{pound} = 4 \text{ ounces}.$$

Extracting the square root of both sides of this expression gives

$$\sqrt{\frac{1}{4} \text{ pound}} = \sqrt{4 \text{ ounces}},$$

or

$$\frac{1}{2} \text{ pound} = 2 \text{ ounces}.$$

Paradox 6.

In attempting to solve the system of two equations in two unknowns:

$$\begin{cases} x + y = 1, \\ x + y = 2, \end{cases}$$

we are forced to the conclusion that, since 1 and 2 are equal to the same thing, they must be equal to each other—that is, $1 = 2$.

Where are the errors? Paradox 1 is too obvious to spend any time on. As a matter of fact, we had to stretch our imagination a little in order to get "equals" out of the statements (1) and (2).

In Paradoxes 2 and 5 we performed the operations of multiplication and root extraction only on the *numbers*, and not on the *units* involved. Our conclusion in Paradox 2, for example, should have been

$$1(\text{pound})^2 = 256(\text{ounces})^2.$$

Now a "square pound" is a rather difficult thing to visualize. It would be clearer if we used feet and inches. Our argument would then run as follows:

$$2 \text{ feet} = 24 \text{ inches;} \tag{1}$$

$$\frac{1}{2} \text{ foot} = 6 \text{ inches.} \tag{2}$$

Therefore 1 square foot = 144 square inches—a result which is evidently correct.

Paradox 3 reminds us rather forcibly of the fact that the axiom concerning "equals divided by equals" carries with it a rider to the effect that the divisors shall not be zero. We shall have more to say about this point before very long.

Paradox 4 recalls another item which may well have been forgotten. In extracting a square root, both the positive and negative signs must be taken into consideration. That is to say, the expression in question yields the two correct identities $+a = +a$ and $-a = -a$. Here is another matter which will receive more attention later on.

Paradox 6 shows us that the axioms cannot be applied blindly to equations which are true only for certain values of the variables, or unknowns. The values of x and y for which both the equations (1) and (2) are true must be taken into account, and there are no values of x and y for which $x + y = 1$ and, at the same time, $x + y = 2$.

Paradoxes in probability

A real number—rational or irrational—between 0 and 10 is chosen at random. What is the probability that it is greater than 5?

⌐...⌐ We divide a segment 10 units long into two intervals, each of length 5 units. Then the probability that the number chosen lies in the favourable interval is $\frac{5}{10}$, or $\frac{1}{2}$.

Let us, for a moment, look at a related problem.

A real number—rational or irrational—between 0 and 100 is chosen at random. What is the probability that it is greater than 25?

This time we divide a segment 100 units long into two intervals, the first of length 25 units, the second of length 75 units. The favourable interval in this case is the second one, and the probability that the random number is greater than 25 is $\frac{75}{100}$, or $\frac{3}{4}$.

Now consider the fact that every number between 0 and 25 has a square root which lies between 0 and 5, and every number between 25 and 100 has a square root which lies between 5 and 10. We can therefore interpret the results of our two problems in this way: if a *number* between 0 and 10 is chosen at random, the probability that it is greater than 5 is $\frac{1}{2}$, whereas if the *square of the number* is chosen at random, the probability that the number is greater than 5 is $\frac{3}{4}$!

What is going on here? Should not the desired probability be the same regardless of whether the number *or* its square is chosen at random? Let us scrutinize the two problems more carefully.

In the first problem we probably based our assumption that the two intervals were equally likely on the idea that the real numbers between 0 and 10 are evenly distributed along the line—that there are, so to speak, just as many real numbers between 0 and 5 as between 5 and 10. But now consider the squares of all such numbers. Every number in the interval 0 to 5 has a square which lies in the interval 0 to 25, and every number in the interval 5 to 10 has a square which lies in the interval 25 to 100. There are, in other words, just as many real numbers between 0 and 25 as between 25 and 100. ⌞...⌟ We are therefore led to the conclusion, whether we like it or not, that the intervals 0 to 25 and 25 to 100 are equally likely to contain a number picked at random between 0 and 100.

But in the second problem we went on the assumption that the numbers between 0 and 100 are distributed evenly along the line and that there are, so to speak, *three times* as many numbers between 25 and 100 as between 0 and 25. That is to say, we assumed that the interval 25 to 100 is *three times* as likely to contain the random point as the interval 0 to 25. This assumption, after all, is a reasonable one. It is the one we should have made had we had no knowledge of the first problem, but had been thinking simply of a number picked at random between 0 and 100.

The way out of all this confusion is not entirely clear. The difficulty is concerned with the proper choice of a set of equally likely cases, a matter on which the mathematicians themselves are not agreed. One group, following Bertrand,° would dismiss all such problems by pointing out that infinity is not a number and that we cannot describe, in terms of finite probabilities, choices made at random from an infinitude of possibilities. This attitude indeed offers a way out, but not a very happy one, for it requires junking many results and techniques which have been found to be extremely useful.

Perhaps the most satisfactory attitude for us to take is the pragmatic one. Granting, when the number of cases is infinite, that the choice of a set of equally likely cases is arbitrary, let us choose that set which common sense tells us is the most practical for the particular problem under consideration. Thus, in the two problems we have been discussing, the set used in the first problem certainly appears to be more practical for that problem than the set used in the second problem would be. What man of the street, confronted with the problem of determining the probability that a random number between 0 and 10 is greater than 5, would go off into calculations concerning the square of the random number and come out with the answer $\frac{3}{4}$? The common-sense answer is $\frac{1}{2}$.

We shall see shortly that the pragmatic attitude is not always entirely satisfactory, but the great argument in favour of it is the status of the theory of probability today. The theory is what it is because those who were responsible for its development were practical men who had the good common sense to make practical assumptions when they needed them. Had they stopped to wrangle over every theoretical point which arose, the theory might have died almost at birth. Instead, it has grown to be a powerful weapon of research in many fields.

Note

Bertrand: Parisian mathematician Joseph Louis François Bertrand (1822–1900), an important teacher and probability theorist.

Fermat's Last Theorem

Hans Rademacher and Otto Toeplitz, 1957

Otto Toeplitz was one of the leading mathematicians of his day, working on infinite linear and quadratic forms; Hans Rademacher worked on analysis and number theory. German by birth, both were removed from their university positions by the Nazis and later emigrated, Toeplitz to Palestine, Rademacher to the United States.

In this book the two authors aimed "to show that the aversion toward mathematics vanishes if only truly mathematical, essential ideas are presented."

Hans Rademacher (1892–1969) and Otto Toeplitz (1881–1940), trans. Herbert Zuckerman, *The Enjoyment of Mathematics: Selections from Mathematics for the Amateur* (Berlin, 1933; Princeton, 1957), pp. 88–95. ©1957 by Princeton University Press. Reprinted by permission of Princeton University Press.

Pythagorean Numbers and Fermat's Theorem

1. According to the Pythagorean theorem, the square on the hypotenuse of a right triangle has the same area as the sum of the squares on the two legs. Conversely, if three line segments are such that the square on one is equal to the sum of the squares on the other two, then the three segments will form a right triangle. The equation $a^2 + b^2 = c^2$ represents the fact that the segments of length a, b, c are the sides of a right triangle.

⌞...⌟ The hypotenuse and legs of an isosceles right triangle are incommensurable, ⌞and⌟ the equation $2a^2 = c^2$ can never be satisfied by whole numbers a and c. Are there any right triangles in which the sides are commensurable?° In other words, can the equation

$$a^2 + b^2 = c^2 \tag{5.1}$$

be satisfied by three whole numbers? A simple and very well known example shows that the answer is yes:

$$3^2 + 4^2 = 5^2 \text{ or } 9 + 16 = 25.$$

Are there other answers? How can we find them? In this chapter we shall find the complete answer to these questions.

2. If we have a solution a, b, c of (5.1), we can easily find another by multiplying each of the terms $a, b,$ and c by any whole number. Since 3, 4, 5 is a solution, we can multiply by 2 to find 6, 8, 10. This gives us

$$6^2 + 8^2 = 10^2.$$

More generally, $3n, 4n, 5n$ will be a solution if n is any whole number. In the same way, if a, b, c is any solution, then an, bn, cn is also a solution, for from $a^2 + b^2 = c^2$ we have $a^2 n^2 + b^2 n^2 = c^2 n^c$ or $(an)^2 + (bn)^2 = (cn)^2$. This way of finding new solutions is trivial and therefore not of much interest. It is more interesting to find the basic solutions, those that can't be found merely by multiplying another solution by some whole number. We shall call such solutions "reduced solutions," those solutions in which a, b, and c do not have a common divisor. Thus 3, 4, 5, is a reduced solution.

If two or more numbers have no common divisor we shall say that they are "relatively prime." In a reduced solution, each pair of the numbers a, b, c is relatively prime. For if a and b, say, had a common divisor d, they would also have every divisor of d as a common divisor. Some prime p would divide d; at worst, d would be p itself. Then a and b would have the common divisor p and we could write

$$a = pa_1, b = pb_1$$

Equation (5.1) would then become

$$p^2(a_1^2 + b_1^2) = c^2,$$

from which we see that p^2 would divide c^2. Then p would divide $c^2 = c.c$, and hence one of the two equal factors c. That is, p would divide c as well as a and b. In the same way, a common prime factor of a and c or of b and c is a common factor of all three numbers.

3. Now we are looking for the solution a, b, c of (5.1), in which every pair of a, b, and c is relatively prime. No two of the numbers can be even, that is, divisible by two; at most, only one can be even. Neither can all three numbers be odd, however. The square of an odd number $a = (2l + 1)$ is $a^2 = 4l^2 + 4l + 1$, which is again odd. Then, if a and b are odd, so are a^2 and b^2, so $a^2 + b^2$ is even and cannot be equal to the square of an odd c.

The only possibility that remains is that two of the numbers a, b, c are odd and one is even. Furthermore, we can see that c must be odd, for if c is even it is divisible by 2 and c is divisible by 4. The other two numbers must be odd,

$$a = sl + 1, b = 2m + 1,$$

and we find

$$a^2 + b^2 = (4l^2 + 4l + 1) + (4m^2 + 4m + 1) = 4(l^2 + l + m^2 + m) + 2.$$

This number is even, but on division by 4 , it leaves the remainder 2 and therefore could not equal c^2, which is divisible by 4.

We are now left with c odd and one of the numbers a and b even, the other odd. We shall let a be the odd number, b the even one. Thus in our example we have $a = 3$, $b = 4$, $c = 5$.

4. Equation (5.1) can be written in the form

$$b^2 = c^2 - a^2 = (c + a)(c - a). \tag{5.2}$$

Here $(c + a)$ and $(c - a)$, being the sum and difference of two odd numbers, are both even. Their only common factor is 2. In other words, $\frac{c+a}{2}$ and $\frac{c-a}{2}$ are relatively prime, as we can see by supposing that d divides these two numbers. Then

$$\frac{c + a}{2} = df, \frac{c - a}{2} = dg,$$

and on adding and subtracting these two equations we find

$$c = d(f + g), a = d(f - g).$$

Then d divides both a and c, and this contradicts our assumption that they are relatively prime.

Since b, $c + a$, and $c - a$ are all even, we can write (5.2) in the form

$$\left(\frac{b}{2}\right)^2 = \frac{c + a}{2} \cdot \frac{c - a}{2}, \tag{5.3}$$

where the fractions are only apparent since each is actually a whole number. This equation expresses the square $\left(\frac{b}{2}\right)^2$ as a product of two *relatively prime* factors $\frac{c+a}{2}$ and $\frac{c-a}{2}$. We now come to the essential step of the proof. We prove that $\frac{c+a}{2}$ and $\frac{c-a}{2}$ must each be squares. If $\frac{b}{2}$ is factored into prime factors,

$$\frac{b}{2} = p^\alpha q^\beta r^\gamma \dots,$$

where $p, q, r, \dots$ are different primes, then we have

$$\left(\frac{b}{2}\right)^2 = p^{2\alpha} q^{2\beta} r^{2\gamma} \dots.$$

All the prime factors of $\left(\frac{b}{2}\right)^2$ must appear in $\frac{c+a}{2}$ and $\frac{c-a}{2}$ taken together. However, each prime factor p must either appear only in $\frac{c+a}{2}$ or only in $\frac{c-a}{2}$, since $\frac{c+a}{2}$ and $\frac{c-a}{2}$ have no common factor. Therefore the prime factors of $\left(\frac{b}{2}\right)^2$ are distributed between $\frac{c+a}{2}$ and $\frac{c-a}{2}$ in such a way that each prime power $p^{2\alpha}, q^{2\beta}, r^{2\gamma}, \ldots$ goes entirely into $\frac{c+a}{2}$ or $\frac{c-a}{2}$. Therefore $\frac{c+a}{2}$ and $\frac{c-a}{2}$ contain only even powers of their prime factors and consequently each is a square.

5. We can now write

$$\frac{c+a}{2} = u^2, \frac{c-a}{2} = v^2, \tag{5.4}$$

$$\left(\frac{b}{2}\right)^2 = u^2 v^2, \tag{5.5}$$

where u and v, like u^2 and v^2, are relatively prime. From (5.5) we have

$$b = 2uv, \tag{5.6}$$

and by addition and subtraction of the equations (5.4) we find

$$c = u^2 + v^2, a = u^2 - v^2. \tag{5.7}$$

Since c and a are both odd, one of the squares u^2 and v^2 must be even and the other odd; in any other case their sum and difference would be even. The same must be true of u and v. We shall say that two such numbers are of "opposite parity."

We have now proved that if a, b, c is a reduced solution of (5.1), then a, b, and c can be represented in the form (5.6) and (5.7) by means of two numbers u and v which are relatively prime and of opposite parity. In our old example, $a = 3, b = 4, c = 5$, we have

$$u^2 = \frac{5+3}{2} = 4, v^2 = \frac{5-3}{2} = 1,$$

$$u = 2, v = 1,$$

and, in agreement with (5.6),

$$b = 4 = 2 \cdot 2 \cdot 1 = 2uv.$$

⌊···⌋

Some examples of Pythagorean numbers derived from (5.6) are given in the following list:

$$u = 2, \quad v = 1: \quad a = 3, \quad b = 4, \quad c = 5$$
$$u = 3, \quad v = 2: \quad a = 5, \quad b = 12, \quad c = 13$$
$$u = 4, \quad v = 1: \quad a = 15, \quad b = 8, \quad c = 17$$
$$u = 4, \quad v = 3: \quad a = 7, \quad b = 24, \quad c = 25$$
$$u = 5, \quad v = c: \quad a = 21, \quad b = 20, \quad c = 29$$
$$u = 5, \quad v = 4: \quad a = 9, \quad b = 40, \quad c = 41$$

A glance at the table shows that not only is b even, but it is always a multiple of 4. This is true because $b = 2uv$, and either u or v is even.

6. Now that we have completely solved the problem connected with equation (5.1), a whole series of generalizations comes to mind. We can ask for a similar discussion of the equation

$$x^3 + y^3 = z^3, \tag{5.8}$$

or of

$$x^4 + y^4 = z^4, \tag{5.9}$$

or, more generally, of

$$x^n + y^n = z^n \tag{5.10}$$

for any $n > 2$. Pierre de Fermat (1601–1665) asserted that the equation has no solution in positive whole numbers x, y, z for $n > 2$. This statement, that has never been proved or disproved, is called Fermat's Theorem, or Fermat's last theorem ̦...̩.° However, the assertion has been proved for certain values of n. For example, it has been proved for all n from 3 to 100 by Kummer (1810–1893) and his followers. Before this, Euler (1707–1783) had proved it for (5.8) and (5.9).

Notes

commensurable: capable of being expressed by a set of whole numbers.
Fermat's last theorem was proved in 1994 by Andrew Wiles (b. 1953).

Where Does It End?

Dan Pedoe, 1958

This discussion of the paradoxical properties of the real numbers picks up the theme, from earlier in this chapter, of Northrop's "paradoxes in probability." Dan Pedoe worked on geometry in various universities in the United Kingdom and United States, as well as in Khartoum and Singapore. One of the things he wished to emphasize in this book was that "not all mathematicians are witty and clever."

Dan Pedoe (1910–1998), *The Gentle Art of Mathematics* (Mineola, NY, 1973), pp. 5–6, 56–63. Reprinted by permission of Dover Publications.

Where does it end?

The infinite appears in a variety of forms in mathematics. ⌞...⌟ Mathematicians, as we have seen, are not afraid of thinking of very large numbers, but any *definite* number, however large, is considered *finite*, not infinite. Thus the population of the world at any instant is a finite number, and so is the number of grains of sand on all the beaches of Britain. On the other hand, the number of natural numbers, 1, 2, 3, 4, ... and so on, is infinite, because, however far one goes in counting, say up to a number N, the number $N+1$ is another natural number which has not yet been allowed for. As we shall see later, the class of natural numbers, or integers, is one of the fundamental measures for infinity.

Another example of an infinite class, or set, of numbers is given by the *squares* of all the natural numbers:

$$1, 4, 9, 16, 25, 36, 49, \ldots\ldots\ldots$$

That this is an infinite class is proved in the same way. If we hopefully arrive at the number n^2, we cannot stop there, because there is also a number $(n+1)^2$ in the class. Hence we have two classes, each containing an infinity of natural numbers, and it is natural to enquire whether one of the classes contains "more" numbers than the other.

It is certainly true that all members of the second class are *contained* in the first, since n^2, which is a typical member of the second class, is also an integer, and therefore a member of the first class. Again, there are many members of the first class which are not perfect squares, and are therefore not members of the second class. Can we not therefore say that, in spite of the fact that both classes contain an infinite number of members, the first class somehow contains a *greater infinity* than the second?

This very problem was discussed by Galileo in his *Dialogues.*° He came to the conclusion that all that we can say about the two classes is that each of them is infinite, and that the relations of equality and inequality can be applied to finite, but not to infinite classes. There the matter rested until the possibility of comparing *degrees of infinity* was realised by Cantor, a German mathematician born in Russia, in 1873. Out of his work a most astonishing branch of mathematics has developed. The fundamental ideas are extremely simple.

In order to understand Cantor's reasoning, we must begin with *counting.* What do we mean when we say that there are twenty-one members in a finite set of objects? It is not enough to answer that we point at each member of the class in turn, and recite "One, two, three, ..., twenty, twenty-one." The ability to perform this operation indicates the possession of a highly specialised vocabulary of number words. How could we convey the sense of twenty-one objects to someone who does not understand our language, and whose own language is so undeveloped that there are no words for "five," "six," ..., the only number words being "one," "two," "three" and "four," anything above "four" being many?

We could obviously convey the sense of "twenty-one" by cutting this number of notches on a stick, or by depositing this number of beads on the ground, or by holding up both hands with all the fingers spread out twice in succession, and then holding up one solitary finger. All these methods will be familiar to the reader from books he has read, or from his own experience. There is no doubt that the *sense* of the number twenty-one can be conveyed.

What we have done, in technical language, is to set up a "one-to-one" correspondence between the marks cut on a stick, or the beads, or our fingers, and the objects whose *number* is twenty-one. This is counting in its simplest and most fundamental form. For every object there is *one* notch on the stick, *and* every notch on the stick corresponds to just *one* object.

We consider another example of one-to-one correspondence. I know that the room I teach in contains twenty-one desks. If, when I enter the room, I see that all the desks are occupied, then I know that twenty-one students have decided to attend that particular class. If, on the other hand, some of the desks are empty, then fewer than twenty-one students are present. If, finally, all the desks were occupied, and some students were standing, I should know that more than twenty-one students had turned up, and I should need more desks.

Hence a one-to-one correspondence between desks and students indicates the *same* number of desks and students. If there are seats to which no students belong, there is a *larger* number of desks than students. But if all the desks are occupied, and some students have no desks assigned to them, then the number of students is *greater* than the number of desks.

This is all very obvious, but it contains the germ of Cantor's great idea. The *actual number* of desks does not enter into the notion of equality or inequality. Hence we say:

If two infinite classes are such that a one-to-one correspondence can be set up between their members, then the two classes have the *same transfinite number* of members. This defines equality.

If a one-to-one correspondence can be set up between the members of an infinite class x and *a proper subset y* of an infinite class z (the class y is contained in z and does not coincide with it), then the class z is said to have *greater* transfinite number than the class x.

With these new concepts in mind, we return to the infinite class which consists of the natural numbers

$$1, 2, 3, 4, 5, 6, 7, \ldots \ldots$$

We can set up a one-to-one correspondence between the natural numbers and their squares

$$1, 4, 9, 16, 25, 36, 49, \ldots$$

by making any number n in the first class correspond to n^2 in the second class, and making any number n^2 in the second class correspond to n in the first class, thus:

1	2	3	4	5	6	7	.... n	....
1	4	9	16	25	36	49	.... n^2	....

It is true that we cannot demonstrate this correspondence for *every* member of each class, but it is sufficient to demonstrate it for a *typical* member n of the first class, and a typical member n^2 of the second class.

Having demonstrated the existence of a one-to-one correspondence, we can conclude that the class of the squares of all the natural numbers has the same transfinite number as the class of all the natural numbers! This result is not what might have been anticipated, seeing that the second class is a proper subset of the first.

Similarly, the class of all *even* numbers has the same transfinite number as the class of all natural numbers. The one-to-one correspondence looks like this:

$$1 \quad 2 \quad 3 \quad 4 \quad 5 \quad 6 \quad 7 \quad \ldots\ldots\ldots n \quad \ldots\ldots\ldots$$
$$2 \quad 4 \quad 6 \quad 8 \quad 10 \quad 12 \quad 14 \quad \ldots\ldots\ldots 2n \quad \ldots\ldots\ldots$$

Again, the class of all *odd* numbers has the same transfinite number as the class of natural numbers, the one-to-one correspondence being

$$1 \quad 2 \quad 3 \quad 4 \quad 5 \quad 6 \quad 7 \quad \ldots\ldots\ldots n \quad \ldots\ldots\ldots$$
$$1 \quad 3 \quad 5 \quad 7 \quad 9 \quad 11 \quad 13 \quad \ldots\ldots 2n - 1 \quad \ldots\ldots\ldots$$

In each of these three examples the class of natural numbers has been put in one-to-one correspondence with a *part of itself*. In other words, we have been demonstrating that *the whole is equal to part of itself.* This is in direct contradiction to the familiar assumption, or axiom, first encountered in geometry, that *the whole is equal to the sum of its parts, and is therefore greater than any of them.* This axiom, of course, refers to *finite* magnitudes.

That the whole is equal to part of itself, paradoxical as it may seem, is a conclusion which involves the essence of infinite magnitude. We have not, the reader may have noticed, defined an *infinite class* as yet. But we can now, with Cantor, *define* an infinite class as a class which can be put into one-to-one correspondence with a part (or proper subset) of itself!

If the reader feels that this definition is unnecessarily abstract, and that it is always possible, by *counting* the members, to see whether a class is infinite or not, he must be warned that we shall soon produce a class whose members *cannot be counted*! But before we produce this specimen, we show that the class of all positive rational numbers p/q, where p and q are integers, *can* be counted. This means that the class has the same transfinite number as that of the natural numbers.

It is high time that we gave this last transfinite number a name, and we cannot do better than to follow Cantor, and to use the first letter *aleph* $\aleph$ of the Hebrew alphabet (Cantor actually used $\aleph_0$) to denote the transfinite number of the class of natural numbers. Classes which have the transfinite number $\aleph$ are said to be *countable*, or *denumerable*.

It is surprising that the class of all positive rational numbers turns out to be denumerable, because we can interpose an infinity of rational numbers between any two given rational numbers. For instance, between 0 and 1 we can interpose the rationals

$$\frac{1}{2}, \frac{2}{3}, \frac{3}{4}, \frac{4}{5}, \ldots\ldots, n/(n+1), \ldots;$$

between 0 and $\frac{1}{2}$ we can interpose

$$\frac{1}{3}, \frac{2}{5}, \frac{3}{7}, \frac{4}{9}, \ldots\ldots, n/(2n+1), \ldots;$$

and so on. Because of this property we might well expect that the transfinite number of the class of all positive rationals would exceed $\aleph$. But Cantor showed that the positive rationals can be counted, and we now give his proof.

⌊(The proof follows. Then ....)⌋

We prove that there is an infinite class which has a transfinite number greater than $\aleph$, and is therefore not denumerable. This class is the class of all *real numbers* between 0 and 1. From our point of view a real number is one which has a decimal expansion, so that the class of real numbers between 0 and 1 are all those that are of the form

$$0.a_1 a_2 a_3 \ldots\ldots\ldots,$$

where the sequence of digits after the decimal point may terminate or be infinite.

To prove that the class of real numbers lying between 0 and 1 is not denumerable we assume the contrary, and show that this leads to a contradiction. We assume, then, that it is possible to establish a one-to-one correspondence between the real numbers between 0 and 1 and the natural integers. We shall then show that there is a real number which has not been counted, and has no place in the enumeration.

Before we begin the proof we prepare our real numbers. ⌊...⌋ A number has a *unique* decimal expansion unless the expansion terminates after a finite number of decimal places, when it may also be terminated by an an infinite decimal expansion, involving the sequence .999999.... For example, the number .345 may also be represented by the decimal .3449999....

This is because $0.99\ldots = 1$. If therefore a number has an infinite decimal expansion, no other infinite decimal expansion can represent the same number. A difference in any one digit in two infinite decimal expansions means that the expansions represent two distinct numbers. We wish to use this fundamental property of decimal expansions, and therefore we see to it that all numbers to be considered are clad in an infinite decimal expansion.

The assumption is that all decimals of the form $.a_1a_2a_3\ldots$ can be put in order. Let this order be as shown below:

1	.	a_1	a_2	a_3	a_4	a_5	a_6	a_7	a_8	a_9	......
2	.	b_1	b_2	b_3	b_4	b_5	b_6	b_7	b_8	b_9	......
3	.	c_1	c_2	c_3	c_4	c_5	c_6	c_7	c_8	c_9	......
4	.	d_1	d_2	d_3	d_4	d_5	d_6	d_7	d_8	d_9	......
5	.	e_1	e_2	e_3	e_4	e_5	e_6	e_7	e_8	e_9	......
6	.	f_1	f_2	f_3	f_4	f_5	f_6	f_7	f_8	f_9	......
.	.	.	.	.	.	.	.	.	.	.	......

We assume, of course, that the numbers on the right are all known, and that we have a list which extends as far as we wish it to extend. We now show that there is a decimal lying between 0 and 1 which *does not appear anywhere in the list*. This decimal is *constructed* as follows; we shall call it $.z_1z_2z_3z_4\ldots$.

If a_1 is any one of the digits 0, 1, 2, 3, 4, 5, 6, 7, then z_1 shall be 8. If a_1 is either 8 or 9, then z_1 shall be 1.

If b_2 is any one of the digits $0, 1, \ldots, 7$, then z_2 shall be 8. If b_2 is either 8 or 9, then z_2 shall be 1.

If c_3 is any one of the digits $0, 1, 2, \ldots, 7$, then z_3 shall be 8. If c_3 is either 8 or 9, then z_2 shall be 1.

The procedure should now be quite clear. This decimal we have written down lies between 0 and 1, but where is its place in our enumeration? It cannot be the first decimal, since by construction z_1 differs from a_1. It cannot be the second decimal, since by construction z_2 differs from b_2. It cannot be the third decimal, since by construction z_3 differs from $c_3, \ldots$. It cannot be the nth decimal since by construction z_n differs from the nth digit of the decimal in the nth place in the enumeration.

Hence the decimal we have constructed has no place in the enumeration, and the claim that all decimals lying between 0 and 1 have been put in order must be false. This proves that there is an infinite class of numbers whose members, in the ordinary sense, cannot be counted. This class contains, as a proper subset, the class of rationals

$$\frac{1}{1} \quad \frac{1}{2} \quad \frac{1}{3} \quad \frac{1}{4} \quad \frac{1}{5} \cdots\cdots \frac{1}{n} \cdots$$

which evidently has the transfinite number $\aleph$. We are therefore justified in saying that the transfinite number of the class of real numbers lying between 0 and 1 is *greater* than $\aleph$.

Is there an infinite class whose transfinite number lies *between* $\aleph$ and the transfinite number of the class of real numbers lying between 0 and 1? It is *believed* that there is not, but this has not yet been proved, and must be regarded as one of the great unsolved problems of modern mathematics.

Note

Dialogues: Galileo Galilei, *Discourses and Mathematical Demonstrations Relating to Two New Sciences* (Leiden, 1638).

Yamátárájabhánasalagám

Sherman K. Stein, 1963

Sherman K. Stein has written books on mathematics in everyday life and on Archimedes, as well as textbooks; he spent most of his career at the University of California, working in in algebra, combinatorics, and pedagogy.

This piece on combinatorics illustrates a subject which featured in many mathematics popularizations and recreations, and Stein's approach to it also connects with Rouse Ball's discussion of route problems earlier in this chapter.

Sherman K. Stein, *Mathematics: The Man-made Universe: An Introduction to the Spirit of Mathematics* (San Francisco and London, 1963), pp. 110–118.

Memory Wheels

Mathematics, like every branch of knowledge, is the product of the interplay between past and present, between accumulated knowledge and curiosity, between an autonomous structure and the tastes and needs of the time. What one age considers a pressing question, another may not ask at all. The pure mathematics of one era may be applied in another, perhaps centuries later. Pushing into the unknown, the mathematician is an explorer who is likely to find what he did not seek and who cannot predict how others will use his discoveries.

A problem on which I once worked illustrates these aspects of the growth of knowledge, and its solution captures the flavor of mathematical research. This particular adventure began when the composer George Perle told me about an elaborate theory of rhythm that had been developed in India perhaps some thousand years ago. "While reading about this theory," he said, "I learned my one and only Sanskrit word: *yamátárájabhánasalagám*." I asked him what it meant.

"It's just a nonsense word invented as an aid in remembering the names of certain rhythms."

"If a person can remember that," I replied, "he can remember anything."

"There is a lot in those ten syllables," said Perle. "As you pronounce the word you sweep out all possible triplets of short and long beats. The first three syllables, *ya má tá*, have the rhythm short, long, long. The second through the fourth are *má tá rá*: long, long, long. Then you have *tá rá ja*: long, long, short. Next there are *rá ja bhá*: long, short, long. And so on."

I wrote down the word and saw that what Perle said is true. Each successive triplet of syllables displays a different pattern, and the whole word displays all eight possible patterns, giving each once and only once. As a mathematician I was fascinated to find that such a sequence could exist.

That night I returned to the ancient word. To strip it of irrelevancies I replaced the syllables with digits, letting 0 stand for a short beat and 1 for a long beat. In this notation *yamátárájabhánasalagám* became 0111010001.

After staring at the simplified string for a while, I noticed a lovely thing. The first two digits are the same as the last two; if I bent the string into a loop it would look like a snake swallowing its own tail. That is, the last 01 could be placed over the first 01, and the two pairs of digits would merge into a single pair. Instead of a line of ten digits I now saw a circle of eight.

I could begin anywhere on this "memory wheel" and move around it in either direction, sweeping out triplets of 0's and 1's. Starting at the top and reading counterclockwise, for example, gave 011, 111, 110, 101, 010, 100, 000, and 001.

The next thing that occurred to me, as it would automatically to any mathematician, was to generalize what I had found. Is there a "word" for listing all quadruplets of 0's and 1's once and only once? For quintuplets? For groups of any size? And if so, does the snake always swallow its tail?

Before attacking the problem for quadruplets I decided to go back and look at couplets. Is there a word that lists each of the four couplets 00, 01, 10, and 11 exactly once? Does it close up on itself? Writing down the sequence 0011, I saw that I already had the three couplets 00, 01, and 11. Adding one more 0, to form 00110, gave the final couplet: 10. I noted that, since the first and last digits are 0's, the snake does swallow its tail. The five-digit word could be bent into a four-digit wheel containing each couplet once and only once.

Now I was ready to take on the quadruplets. I began methodically by listing all the possible groups of four digits composed of 0's and 1's. To do this I first wrote down the eight triplets and placed a 0 in front of each. This gave me all the quadruplets beginning with 0. Then I repeated the triplets and placed a 1 before each, obtaining the quadruplets beginning with 1.

0000	1000
0001	1001
0010	1010
0011	1011
0100	1100
0101	1101
0110	1110
0111	1111

Since the list contained 16 quadruplets, I saw that any memory word that would sweep out each of them once (if any exists) would have to have 19 digits. The first four digits make one quadruplet, and each of the next 15 numbers completes another.

Somewhere in the word I was seeking, I knew there would have to be a string of four 1's and, somewhere, four 0's. Why not put them together and see what would happen? I wrote down the eight symbols 11110000. Then I checked off my list the five quadruplets it contained: 1111, 1110, 1100, 1000, and 0000. So far so good. To avoid getting 0000 twice, I next had to add a 1, which gave 111100001. I checked off 0001 on the list. Adding 010

produced three more quadruplets, all new, and brought the sequence to
111100001010.

From here I proceeded one digit at a time, checking the list to make
sure there were no duplications. At each of the next three positions
the choice was clear: one of the digits would form a quadruplet already
checked; the other would not. I had then reached the 15-symbol word:
111100001010011.

Only four to go. Considering the next digit, I found that neither 1
nor 0 provided a duplication. But a 1 would lead to trouble in the next
position, where either 0 or 1 would produce a duplication. I was afraid I
might not be able to reach 19 symbols after all. It turned out, however,
that a 0 in the sixteenth position involved no such difficulty, and the last
three positions presented unambiguous choices. I had found my word of
19 symbols—a word containing each of the 16 quadruplets precisely once:
1111000010100110111.

As soon as I had finished, I looked at the first three symbols and the
last three. They were the same. This snake, too, could swallow its tail. This
19-symbol word could be bent into a wheel of 16 symbols.

Inspired by this success, I decided that there must be memory words for
quintuplets, sextuplets, and so on. Furthermore, I felt sure they would all
close up into wheels. It was time to stop experimenting, however, and look
for a proof of the conjecture.

Grappling with the problem, I began to look at the Indian memory word
in a slightly different way, concentrating on the eight overlapping triplets
it contains:

$$
\begin{array}{ccccccc}
0 & 1 & 1 & & & & \\
 & 1 & 1 & 1 & & & \\
 & & 1 & 1 & 0 & & \\
 & & & 1 & 0 & 1 & \\
 & & & & 0 & 1 & 0 \\
 & & & & & 1 & 0 & 0 \\
 & & & & & & 0 & 0 & 0 \\
 & & & & & & & 0 & 0 & 1 \\
\end{array}
$$

In this light, the word appeared as a means of arranging the triplets such
that the last two symbols of one are the same as the first two of the next.
Suppose the word had not been invented. How would one have gone about

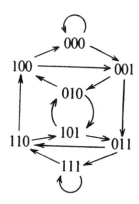

Figure 5.2. A simple pattern of triplets and arrows. (Stein, p. 113.)

finding it? I decided to spread the triplets over a piece of paper and connect the appropriate pairs with arrows. That is, I drew an arrow from one triplet to another whenever the last two symbols of the former were the same as the first two symbols of the latter: for example, 001 $\longrightarrow$ 010 and 001 $\longrightarrow$ 011. After moving the triplets and arrows around a little to make a simple pattern, I obtained this diagram (see Figure 5.2).

As I gazed at the configuration, I suddenly saw it as a map in which the arrows were one-way roads and the triplets were towns. The problem of arranging the eight triplets into a memory word could now be stated in terms of a traveling salesman ﹏, who is looking for a tour over the one-way roads that will take him through each town just once. With the help of the Indian memory word, I traced one possible route. As this illustration shows, the "town" in which the journey ends is adjacent to the one in which it starts (see Figure 5.3).

There is a section of road that will take the salesman from the finishing point back to the start. This, of course, reflects the fact that the memory word closes into a wheel.

Clearly the same scheme would apply to overlapping couplets, to quadruplets, or to groups of any size. The question "Is there always a memory word?" now read "Is there always a route for the salesman?" The question "Does every memory word close up on itself?" now read "Does the salesman always finish his trip in a town adjacent to the town in which his trip began?"

Unhappily, the translation did me no good.

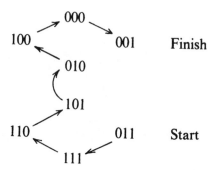

Figure 5.3. The "town" in which the journey ends is adjacent to the one in which it starts. (Stein, p. 114.)

Saddles and Soap Bubbles

Iakov Isaevich Khurgin, 1974

Iakov Isaevich Khurgin published in applied mathematics, including two books which were translated into German; this popular book seems to have been his only appearance in English, though. Here we have two extracts from his section on curved surfaces, dealing with the attractive subjects of mountains and soap bubbles.

Iakov Isaevich Khurgin (dates unknown), trans. Georgy Yankovsky, *Did You Say Mathematics?* (Moscow, 1974), pp. 88–96, 110–113.

The Mathematics of a Saddle

Imagine a mountainous landscape with peaks and slopes and valleys and hills, and passes. It may not sound romantic, but such a surface can be represented analytically by writing

$$z = f(x, y)$$

where z is the vertical coordinate and x and y are coordinates in the horizontal plane. The peaks correspond to maximum values of the function $z = f(x, y)$ and the valleys correspond to minimum values. If you are on a peak, there is only one way of going in any direction, and that is down; if you are in a valley, you can only go up. These points of maximum and minimum on surfaces will soon be of particular interest to us. If you are at some ordinary point of a surface, you can either go up or down. You can even choose a path that remains constantly at the same altitude. Such pathways are obtained by cutting the surface by a horizontal plane. Projections of such pathways onto one common horizontal plane are termed level lines. Those are the lines one sees on maps indicating height above sea level.

An ellipsoid is a figure obtained by rotating an ellipse about its axis of symmetry. An ellipse has two such axes, one major and one minor. Rotation about the major axis yields an elongated, or prolate, ellipsoid, which looks like a cucumber, while rotation about the minor axis yields a compressed, or oblate, ellipsoid, which resembles a ball compressed from both sides.

We choose an arbitrary point P on the surface of an ellipsoid. It is always possible to intersect the ellipsoid with a plane so as to cut off a cap—the portion containing the point P. It is also possible then to choose the cutting plane so that the dimensions of the "cap" are very small (the mathematician would say: less than any preassigned number). Now let us take some point P on an arbitrary surface. If it is possible in any neighbourhood of this point to cut off a "cap" by means of a plane, then we will call this an *elliptic point*. By far not all points of a surface turn out to be elliptic. This will soon become apparent. We could also give a different definition of an elliptic point. We will draw various planes through P. If among these planes there are such that the entire piece of surface in the neighbourhood of the point P lies to one side of the plane, then P is an elliptic point.

Now let us come back to our mountainous terrain. Besides peaks and valleys we are particularly interested in mountain passes. A mountain pass resembles an ordinary horse saddle. Let us take two points A and B on different slopes of a pass (see Figure 5.4). One can travel from A to B along different routes (they are indicated by dashed lines in the figure) each one of which has a highest point denoted by an open circle. Quite obviously, among all these routes from A to B we can choose the highest point that lies lowest. This route is depicted by the boldface dashed line.

Similarly, each solid-line route from point C to point D has a lowest point denoted by an open circle. From among all possible routes from C to D we choose the route whose lowest point is higher than all others. This route is depicted by the boldface solid line.

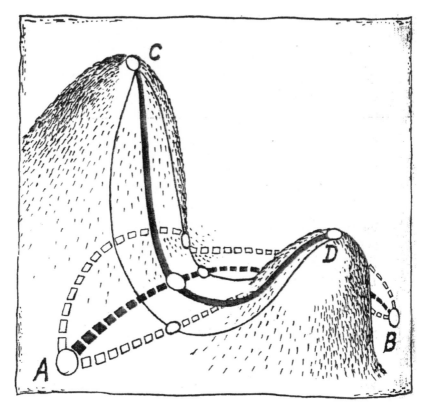

Figure 5.4. A mathematical mountain pass. (Khurgin, p. 92.)

The highest point on the boldface dashed route and the lowest point on the boldface solid-line route coincide. We will call these points saddle points. If we incline the surface slightly, there will be a different saddle point.

We can give a different and perhaps more pictorial description of saddle points. Note that there is no plane that can cut off a "cap" in the neighbourhood of a saddle point. If we pass various planes through a saddle point, then in contrast to the elliptic points, the plane in the neighbourhood of a saddle point will always intersect the surface so that there will be parts of the surface on both sides of the plane. In this description it will be seen that the point will be a saddle point irrespective of any inclination of the surface, or, in other words, irrespective of the choice of directions of axes of the Cartesian rectangular coordinates in space.

Just as a mountainous region can have several mountain passes, so a surface can have several saddle points.

Now I have a question for the reader. Can there be a very large number of saddle points on a surface? For instance, can a surface consist entirely of saddle points? If the answer is negative, then can a bounded piece of surface consist entirely of saddle points?

Before going on to what follows, try to imagine the situation we now have.

The answer is very simple. Take a look at the neck of an ordinary bottle. All its points are saddle points. It is not difficult now to imagine an infinite surface all points of which are saddle points. To do so, take for instance a hyperbola whose equation is $x^2 - y^2 = 1$ and rotate it about the vertical axis. The resulting surface—a hyperboloid of revolution—will consist entirely of saddle points. The hyperboloid is the most elementary surface possessing these properties. For this reason saddle points are also called *hyperbolic points*. Surfaces consisting entirely of saddle points play an important part in our lives.

Take for example the flat diaphragm of an ordinary telephone receiver. Clamp the edge of the diaphragm at several points and suspend small loads at certain other points. After the inevitable oscillations damp out, the diaphragm will assume a position in which all its points wil be saddle points. It is not always possible of course to see this, but that is what the exact theory calls for, namely: under any deformation of the *edge* of a plane membrane (diaphragm), all its interior points will be saddle points.

If various parts of the edge of a membrane are heated in diverse ways, and the heat fluxes are held constant, the temperatures of its points will at first vary, but then will reach a steady state with the influx of heat equal to the efflux. If we lay off the temperatures on a vertical axis, and position the membrane in the horizontal plane, then the appropriate "temperature surface" will consist entirely of saddle points.

The study of surfaces consisting solely of saddle points is closely connected with hydrodynamics, electrostatics, and other very important branches of science.

The shape of a fixed membrane is given by the solution of Laplace's differential equation. The very same equation describes a steady-state irrotational flow of a noncompressible fluid and an established flow of heat, the distribution of forces in an electrostatic field, and a steady-state electric current, the diffusion of salt dissolved in water, and many other phenomena and processes. And all the functions—the solutions of these equations in a geometrical representation—prove to be surfaces that consist entirely of saddle points. That is why the investigation of such surfaces is very essential in a great diversity of fields of physics and technology.

Soap bubbles

The calculus of variations provides us with the machinery for solving a broad range of problems. It is not only used to find the shortest route between point *A* and point *B* but also to solve problems involving the search for a great diversity of extremal quantities.

It is common knowledge that, in a plane, of all figures having a boundary of a given length (or with a given perimeter, as we would say in elementary geometry), the circle has the largest area. In three-dimensional space, the solid of greatest volume for a given area of the bounding surface is a sphere. Conversely, of all solids of a given volume, the sphere has the least surface area. That is the precise reason why soap bubbles appear in the form of spheres.

Let us take up some less obvious problems.

A circle can be the boundary of a surface, say of a pail. Now, of all surfaces having such a boundary, the one with a minimal area is the plane disk stretched over that circle. Now distort the circle so that the curve can no longer be superimposed on the plane. There are a number of surfaces having such a boundary. But how does one find the minimum-area surface among them? That is already a difficult problem, and to solve it analytically requires applying methods of the calculus of variations. It turns out—Euler established this fact—that at every point such a minimal surface is a saddle-like surface.

It is interesting to examine a physical solution of this problem. Put a closed contour (circuit) made of thin tin wire into soapy water. The resulting soapsuds have a small surface tension. A soapy film will adhere to the contour and its area will be the smallest possible area. We have of course disregarded the force of gravity and other forces that prevent the film from attaining a state of stable equilibrium. Stable equilibrium is attained when the area of the film is minimal, since in that case the potential energy due to the surface tension is minimal.

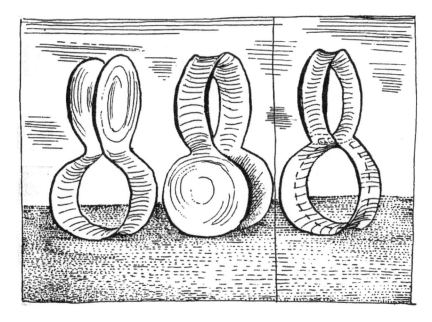

Figure 5.5. Soap bubble topology. (Khurgin, p. 113.)

You have probably forgotten the fun you once had making soap bubbles. Try it again. Take off a few minutes to return to childhood and we'll perform a number of experiments.

Solder a soft wire into a circle with two handles (that'll make it easy to distort into a variety of shapes) and dip it into soapsuds. A soapy film will stretch out around the contour. Now gradually twist it. You will see that by continuous twisting of the contour you can transform a two-sided membrane stretched onto a circle into a one-sided Moebius strip. This is a remarkable fact, for the original surface and the resulting surface are not topological equivalents!

If you bend the circle into a space curve, as shown in Figure 5.5a, then you can stretch three different minimal surfaces on your contour. On the latter (Figure 5.5b), it is possible to draw a closed curve, like the dashed line in the figure, which cannot be contracted continuously into a point without tearing. The other two surfaces do not possess this property. ....

All these beautiful geometrical figures are not only for fun. Surfaces of minimal area are the most rigid, and they find extensive application in the development of rigid structures in engineering practice.

"The Monster" Unveiled

The Times, 1980

The classification of the finite simple groups, described in this extract, was one of the major mathematical achievements of the twentieth century, the more so for involving the efforts of a very large number of individual mathematicians. Historically the subject owed something both to attempts to solve polynomial equations of higher and higher degree and to combinatorial and probability problems. Here we see the London *Times* reporting on the final step in the long process, the construction of the "Monster."

Science report. Mathematics: "The Monster" unveiled

From Clive Cookson of The Times Higher Education Supplement, Washington

One of the great intellectual challenges of pure mathematics, the classification of finite simple groups, has apparently been completed. The final step was taken by Dr Robert Griess, of Michigan University, who constructed a group known informally as The Monster (because its elements total about a billion billion billion billion billion billion, that is, 10 to the power of 54).

Groups provide modern mathematics with one of its most useful tools. They are algebraic structures, consisting of elements of unspecified character which can be combined under a specified operation according to certain rules. Group theory is essential for many scientific fields, particularly crystallography.

The building blocks from which all other groups can be constructed are called "simple groups." Over the past decade many of the most brilliant mathematicians in Europe and the United States have been working to construct and classify finite simple groups. Most simple groups belong to two broad families, but there are a number of exceptions that do not fit in, called "sporadic groups."

By the beginning of this year, 24 sporadic groups had been constructed, and only two more were believed to exist: The Monster (known officially as F1) and a much smaller group called J4. The latter, which has about a hundred billion billion elements, was finally constructed in February by a British team at Cambridge University using a computer program.

In January Dr Griess circulated a note within the mathematical community announcing that he had constructed The Monster without using a computer. But he tantalized his colleagues by refusing to give any details until this month, when he explained his construction in lectures at Chicago University and at the Institute for Advanced Study, Princeton.

The talks were worth waiting for. "It was a tour de force," Dr Jonathan Alperin, of Chicago University said. Dr Griess's construction was exceedingly complicated, but essentially he represented F1 as a group of symmetries in a space of 196,884 dimensions.

The Monster not only fills in the last blank space in the classification of simple groups but also turns out to be a fascinating mathematical construction on its own. At least 19 and possibly as many as 23 of the remaining 25 sporadic groups are embedded as subgroups within it.

More mysteriously, some of The Monster's properties seem to link it to a completely different branch of mathematics, analytical number theory. That connexion could be an extremely fruitful line of future research, some mathematicians feel.

The construction of all 26 sporadic groups completes the main phase of the international effort to classify finite simple groups, an achievement that one senior mathematician compared to the completion of the periodic table of elements in chemistry.

But many loose ends will have to be tied up before it can be stated with absolute confidence that no more sporadic groups remain to be discovered.

According to Dr Alperin, the final proof is likely to be 10,000 pages long and to rely on the contributions of hundreds of mathematicians.

6

"To Ease and Expedite the Work": Mathematical Instruments and How to Use Them

NO ONE WANTS TO READ EXCERPTS FROM AN OLD CALCULATOR MANUAL. YET instruments have been important throughout the history of mathematics, and a very important and prominent dimension of the mathematics writing of the past is writing about mathematical instruments of one kind or another: making them, using them, where to buy them or have them repaired. It would be a pity to miss out on this material, but of course it presents the obvious problem that quite a lot of it is simply incomprehensible unless you have the instrument in question in your hands.

So this chapter is a compromise. It includes sections on the *construction* of several classic types of instruments: compasses, sundials, telescopes, "Napier's bones," and the "nocturnal" or star clock. It has somewhat less about their *use*. I hope that in this way you'll learn something worthwhile about what these instruments were (and are) like, without being faced with material that is impossible to understand.

Today's mathematical instruments are arguably calculators and computers (and here there's no chance of giving a passage about how to make one), or, indeed, computer software. As a representative of this kind of modern mathematical "instrument," I have included a short passage from Peter Duffett-Smith's *Easy PC Astronomy*, illustrating the way that these instruments allow us to perform with ease calculations which by more traditional means would be slow, complex, and inaccurate to the point of infeasibility.

"Cards for the Sea"

Martín Cortés, 1561

Martín Cortés (1532–1589) was the son of Hernán Cortés, the first of the conquistadors. He published his account of navigation (*Breve compendio de la sphera y de la arte de navegar . . .*) in Spanish in Seville in 1551, and a decade later it received this English translation by Richard Eden. Eden (c. 1520–1576) was the translator of various works on subjects including science and exploration, and a promoter of colonial ideas.

Various instruments and mathematical techniques were described in Cortés' book; here we see a description of one of the most basic: how to draw a compass rose, the starting point for both map making and accurate direction finding at sea.

Martín Cortés, trans. Richard Eden, *The Arte of Nauigation, Conteynyng a compendious description of the Sphere, with the makyng of certen Instrumentes and Rules for Nauigations: and exemplified by many Demonstrations. Wrytten in the Spanyshe tongue by Martin Curtes, And directed to the Emperour Charles the fyfte. Translated out of Spanyshe into Englyshe by Richard Eden.* (n.p., 1561), lvi^r–lvii^r.

Arriving to the end desired (which is navigation, the principal intent why I began this work), I say that Navigation or sailing is none other thing than to journey or voyage by water from one place to another, and is one of the four difficultest things whereof the most wise king hath written. ⌊. . .⌋ These voyages being so difficult, it shall be hard to make the same be understood by words or writing. The best explication or invention that the wits of men have found for the manifesting of this is to give the same painted in a Card, for the draught or making whereof, it shall be requisite to know two things: whereof the one is the right position of places, or placing of countries and coasts; the other is the distances that ⌊are⌋ from one place to another. And so the Card shall have two descriptions. The one that answereth to the position shall be of the winds, which the Mariners call lines or points of the compass; and the other, that answereth to the distances, shall be the drawing and pointing of the coasts of the land and of the Islands compassed with the sea.

To paint the winds or lines, you must take skins of parchment or large paper, of such bigness as you will the Card to be. And in it draw two

‚straight‚ lines with black ink, which in the midst shall cut or divide themselves in right angles, the one according to the length of the Card, which
shall be East and West, and the other North and South. Upon the point
where they cut, make a center, and upon it, give a privy or hid circle which
may occupy in manner the whole Card. This circle some make with lead
‚pencil‚, that it may be easily put out. These two lines divide the circle into
four equal parts. And every part of these shall you divide in the midst with a
prick or punct. Then from one punct to another, draw a right Diametrical
line with black ink: and so shall the circle remain divided with four lines
into eight equal parts, which correspond to the eight winds. In like manner
shall you divide every ‚one‚ of the eight into two equal parts: and every part
of these is called a half wind. Then draw from every punct to his opposite
diametrically, a right line of green or azure. Likewise shall you divide every
half wind in the circle, into two equal parts. And from these puncts which
divide the quarters, you shall draw certain right lines with red ink, which
also shall pass by the center, which they call the mother compass or chief
compass of the Card, being in the midst thereof. And so shall come forth
from the center to the circumference 32 lines, which signify the 32 winds.

Beside these said lines, you shall make other‚s‚ equal distant to them,
and of the self same colours, in this manner. From the points of the winds
and half winds that pass by the center, draw certain right lines that pass not
by the center, but be equally divided to those that pass by the center, and of
the same colours and equidistance as are they that pass by the center. And
as these lines concur together as well in the center as in the points of the
winds and half winds that are in the circumference of the circle, they shall
leave or make there ‚an‚other 16 compasses, every one with his 32 winds.
And if the Card be very great, because the lines may not go far in sunder, if
you will make there ‚an‚other 16 compasses, you must make them between
the one and the other of the first 16 points, where the quarters are made
with their winds, as we have said.

It is the custom for the most part to paint upon the center of these
compasses a flower or a rose, with diverse colours and gold, differencing
the lines, and marking them with letters and other marks: especially signing
the north with a fleur de lys, and the East with a cross. This, beside the
distinction of the winds, serveth also for the garnishing of the Card. And
this for the most part is done after that the coast is drawn. And thus much
sufficeth for the draught of the winds.

Making a Horizontal Sundial

Thomas Fale, 1593

The making of sundials has been the subject of many, many treatises, and represents one of the oldest and most widespread practical uses of geometry. Our extracts are from one of the earliest such books to be printed in English, by Thomas Fale, a clergyman. It was Fale's only book and was reprinted in 1626 and 1652.

Thomas Fale (1561–after 1604), *Horlogiographia The art of dialling: teaching an easie and perfect way to make all kinds of dials upon any plaine plat howsoeuer placed: vvith the drawing of the twelue signes, and houres vnequall in them all. Whereunto is annexed the making and vse of other dials and instruments, whereby the houre of the day and night is knowne. Of speciall vse and delight not onely for students of the arts mathematicall, but also for diuers artificers, architects, surueyours of buildings, free-Masons and others.* (London, 1593), 4^v–5^v.

The making of a Horizontal or plain lying Sundial.

Your plate being prepared smooth and plain, draw upon it two lines as in the figure following (6.1), the one AB, the other CD, cutting themselves squarewise: that is, making right angles in the point E. Upon which, make the quadrant of any circle from the line EC to the line EA or EB, and write at C the North, at D the South, at A the East, at B the West. And the line CA, which here is the quadrant, being divided into 90 degrees or parts, the elevation of the Pole shall be accounted in it (which in our example is 52 degrees) from C to A; and at the end of this number draw a line from the centre E, which shall be EF, representing the style and artery of the world.

Then draw another line KL by C, or by some other point of the line DC, squarewise, so long as you can, which shall be called the touch line, or line of Contingence. Then measure with your compasses the least distance of the point O and the line EF or the Style. And with the one foot placed in O, which is the point of intersection, and the other extended toward E, where it shall chance to divide or be placed in the line EC, mark that point or centre with the letter G. And draw with your compasses a half circle upon this centre for the equinoctial circle, from H by C to J, whose diameter must be equally distant to the line LK. Then divide this half circle into twelve equal parts.

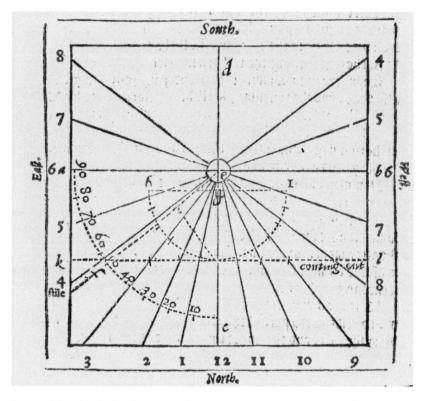

Figure 6.1. The design for a plain lying sundial. (Fale, fol. 5ᵛ. © The Bodleian Libraries, University of Oxford. 4 F 2 Art.BS., p. 5.)

This done, lay a ruler upon the centre *G*, and upon every mark or division made in the half Equator, and where the ruler shall touch the line of contingence, there make marks or pricks, by which pricks draw lines from *E* for the hours. *E C* is the twelfth hour, *E B* the sixth in the morning, *E A* the sixth at evening; the rest you may see in the figure.

And whereas in Summer the fourth and fifth in the morning, and also the seventh and eighth at evening, shall be necessary in this kind of Dial, prolong or draw the lines of four and five at evening, beyond the centre *E*, which shall show the hours of four and five in the morning. And likewise the seven and eight in the morning, for the seven and eight at evening.

You may observe an order both in these and in all other erect direct dials, by dividing the one half of the Equator, drawing hour lines for the forenoon, and observing the same distance from the Meridian line, on the other side for the afternoon; for the line of the eleventh hour in the forenoon is of like distance from the Meridian, that the first is in the afternoon, and the tenth as ‚the‚ second, and so of the rest.

When you would draw or make the half hours, you must divide every part of the Equator into two equal parts, using the ruler and the line of contingence as you find in the drawing of the hour lines.

And this remember for the drawing of the half-hour lines, not only in this kind, but also in all other kinds of dials, which afterward shall follow: the Style must be fixed in the centre E, hanging directly over the Meridian line EC with so great an angle as the lines CEF make, declining from that on neither side.

The Equinoctial circle, the Quadrant, ‚and‚ the line‚s‚ of the Style and of Contingence must be lightly drawn, because they ought to be put out again, in that they serve to no use but for the drawing of the Dial. And this likewise remember in all other kinds of Dials: that the preparative or pricked lines must, after the making of the Dial, be omitted and extinguished, as altogether unprofitable.

This and all other kinds of Dials may most fitly be drawn upon a clean paper, and then with the help of your compasses placed on the plate.

Speaking-Rods

Seth Partridge, 1648

The Scottish mathematician John Napier (1550–1617), better known today as the inventor of logarithms, also devised a system of rapid computation known as "Napier's bones." He first published a description of them in Latin in 1617, a translation followed in 1627, and many later publications elaborated on them. Seth Partridge, the author of this extract, was a mathematics teacher with a particular interest in mathematical instruments; his description of the "speaking rods" was originally written for his students.

Seth Partridge (1603/4–1686), *Rabdologia, or, The Art of numbring by Rods, whereby the tedious operations of Multiplication, and Division, and of Extraction of Roots, both Square and Cubic, are avoided, being for the most part performed by Addition and Subtraction: with Many Examples for the practise of the same: First invented by the Lord Napier, Baron of Marchiston, and since explained, and made usefull for all sorts of men. By Seth Partridge, Surveyor, and Practitioner in the Mathematics.* (London, 1648), pp. 1–7, 22–23.

Speaking-Rods, their Definition and Fabric.

RABDOLOGIA is the Art of counting by Numbering-Rods. As it pleased that ever famous Author the Lord Napier, the first Inventor of that most admirable invention of Logarithms, to call those his excellent tables by the name of Logarithms, that is to say, speaking-Numbers, even so this of the Rods, *Rabdologia*, that is to say, speaking-Rods, or the Speech of Rods. And speaking-Rods they are very properly called, for being so Tabulated, that is to say, placed, set, and laid together, as their nature requireth, they do of themselves tell us, or show forth unto us, with unexpected ease and certainty, and without any operation at all, what the product in any Multiplication is, and what the quotient in any division is, without any charge or trouble of memory, and in Extraction of the Square and Cube root, they do very much ease and expedite the work.

These speaking-Rods may be made either of Silver, Brass, Ivory, or Wood, as the maker and user of them best pleaseth, but they are most ordinarily made of good solid Box⸌wood⸍, and being thereof made, they are as useful as those made of any other substance whatsoever. Nay, I hold them more light and nimble then those made of Metal.

They are foursquare pieces, all placed to a just and even thickness, near about $\frac{1}{5}$ of an inch square, or something better, and their length is just nine times their breadth, so that they be two inches, or near thereupon, in length. The number of these foursquare Rods are most ordinarily Ten, which are enough for ordinary use; you may have 20, 30, or 40 if you will, and as your occasion serveth. To these Rods, how many soever you have, belongeth a broad piece, of the same substance that the Rods be of, called Lamina, and is of the same length and thickness with the Rods, and the breadth is near about the breadth of three of the Rods.

These Rods, being made thus foursquare, have each of them four faces, so that 10 Rods have on them 40 faces, ⸌...⸍ which four faces of every Rod are divided into Nine equal parts, by Nine lines drawn round about the Rod, so that upon each face are Nine perfect squares made. And ⸌all⸍ of those 9 squares or parts are divided into two parts by Diagonal lines. And in those Divisions, or squares, are inscribed Arithmetical figures, beginning at the top, or first square division, and from thence descending downwards

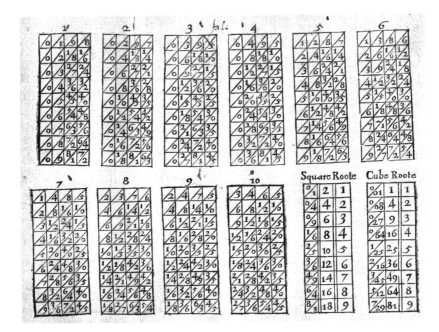

Figure 6.2. The faces of the speaking-rods. (Napier, facing p. 7. © The Bodleian Libraries, University of Oxford. 8° R 84 Med.)

to the Ninth Division in an Arithmetical progression, so that they are by this means fitted ready for many Arithmetical works, as Multiplication, Division, and the extraction of Roots, etc.

But to avoid many words in describing these vocables, behold the shape and figure of all those 10 Rods, and the numbers set on every of their four faces (Figure 6.2). Every practitioner may make them himself by cutting the faces of every one of the printed papers ⌐...⌐, and so placing them⌐ on a square piece of wood as before. Or else they are ready made in Wood, by Master *John Thompson* in Hosier Lane near Smithfield, who makes all kind of Mathematical Instruments, and also by Mr *Anthony Thompson* in Gresham College, and by Mr *Thomas Browne* at the Globe near Aldgate. In Silver or Brass they are made by Mr *Elias Allen*, over against St Clements without Temple Bar.

But note, by the way, that though I have inscribed upon the Rods the same numbers that were used at the first invention, yet I have not inscribed them altogether in the same manner. For there the inscriptions upon the third and fourth faces were set on inverse unto those numbers on the first and second faces. And this way of inscription I find to be the best, because it saveth the labour of turning the Rods end for end, which of necessity must be done when the numbers are set on them inverse, according to that other way.

In this description of the Rods, you may see the figures of each face of the Rod do begin at the top, or upper square space next to the end, and descend down in an Arithmetical progression. As, on the second face of the first Rod, the figures begin at the top with an unit—thus: 1—on the right of the Diagonal line, and descend downward in an Arithmetical progression— thus: 1, 2, 3, 4, 5, 6, 7, 8, 9—every figure downwards increasing by one unit. On the third face of the same Rod the upper figure is 9, and the rest of the figures, descending downwards, being 18, 27, 36, 45, 54, 63, 72, 81, each square exceeding his next by 9. And on the fourth face the figures are 8, 16, 24, 32, 40, 48, 56, 64, 72, every number equally exceeding or increasing his former by 8. But upon the first face of the same first Rod are only ciphers, as there are also upon one of the faces of the second, third, and fourth Rods.

There is also to these Rods a frame to be made, to Tabulate or lay the Rods in when you work with them (you may use them without if you please, but to work with them in a frame is far the better way, and is always intended to be used with them through all this following discourse). This frame is but only a small thin board, as large for length as is the length of the Rods, and as wide as the breadth of all the 10 Rods, with the Lamina, laid all close together side by side; but if you have a greater number of Rods than 10, it is requisite the frame should be the larger, that so you may Tabulate more than 10 places of figures in one number. This board is to have two ledges upon it of equal thickness to the Rods set on squarewise, the one at the top, and the other at the left side, and that ledge on the left side is to be divided into nine equal square parts as the Rods are, and figures set on them, beginning at the head or top square, and descending downwards, thus: 1, 2, 3, 4, 5, 6, 7, 8, 9, like to the second face of the first Rod. It is far better to Tabulate the Rods in a frame, or Table thus made, than without, for the frame keeps them even at the head, and to lie square and close together, and the numbers on the ledge do very readily guide you to the product, in any Multiplication, and to the figure for the quotient, in any Division.

Of Multiplication by the Rods.

Multiplication is either single or compound.

Single Multiplication is when the Multiplicator consisteth but of one single figure only, though the multiplicand consist of many.

Compound Multiplication is when the numbers to be multiplied together, that is to say, the Multiplicand and the Multiplicator, consist, both of them, of more places of figures than one.

For single Multiplication by the Rods, you are first to Tabulate the Multiplicand (that is, set out rods whose top figures form the number required), and then look upon the left hand ledge of the frame for the figure of your Multiplicator, against which figure, you have in the line of squares upon the Rods the product sought for.

Example: if you are to multiply 672 by 4. Tabulate three Rods that have on their top squares the three figures 6, 7 and 2, the three figures of the number given, in order as they ought. First, place the Rod with the figure 6 next the ledge, and the 7 next, and the Rod with the figure 2 last. Which being done, look upon the ledge for the figure 4, against which, upon the fourth line of squares on the Rods, you shall see in the first half-Rhomboids 2, and in the first whole Rhomboids 4 and 2, that is, 6. In the second Rhomboids, 8, and, lastly, in the last and outermost half-Rhomboids, 8. So that ﹏﹏﹏ you must set down the sum of that rank, or line of numbers, thus: 2688, which is the true product of 672 multiplied by 4, which was demanded.

Telescopes Refracting and Reflecting

The Juvenile Encyclopedia, 1800–1801

The Juvenile Library appeared in monthly installments during 1800–1801, and thereafter under the title of *The Juvenile Encyclopedia* until 1803; the collected edition eventually filled six volumes. It promised instruction in a wide range of "useful" subjects, and if that were not enough, its anonymous authors also publicized it through a mildly hysterical prospectus ("...the youthful genius of the whole nation ...") and offered monthly prizes (books, scientific instruments, or both) and the chance for the best "student productions" to appear in print. The telescopes described here were among the prizes.

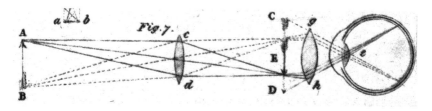

Figure 6.3. The refracting telescope. (*The Juvenile Encyclopedia*, vol. 3, 3rd plate of "Optics." © The Bodleian Libraries, University of Oxford. (Vet.) 3985 e.19–24.)

The Juvenile Encyclopedia. Including a Complete Course of Instruction, on every useful subject: Particularly Natural and Experimental Philosophy, Moral Philosophy, Natural History, Botany, Ancient and Modern History, Biography, Geography and the Manners and Customs of Nations, Ancient and Modern Languages, English Laws, Penmanship, Mathematics, and the Belles Lettres, vol. 3. (London, 1801), pp. 130–132.

What microscopes perform upon minute bodies very near, telescopes perform upon great bodies very remote; namely, they enlarge the angle in the eye under which the bodies are seen, and thus, by making them very large, they make them appear very near; the only difference is that in the microscope the focus of the glasses is adapted to the inspection of bodies very near, in the telescope to such as are very remote. Suppose a distant object at AB (see Figure 6.3), its rays come nearly parallel, and fall upon the convex glass cd; through this they will converge in points, and form the object E at their focus.

But it is usually so contrived, that this focus is also the focus of the other convex glass of the tube. The rays of each pencil,° therefore, will now diverge before they strike this glass, and will go through it parallel, but the pencils all together will cross in its focus on the other side, as at e, and, the pupil of the eye being in this focus, the image will be viewed through the glass, under the angle geb, so that the object will seem at E under the angle DeC.

This telescope inverts the image, and therefore is only proper for viewing such bodies as it is immaterial in what position they appear, as ⸤…⸥ the fixed stars, etc. By adding two convex glasses, the image may be seen upright. The magnifying power of this telescope is found by dividing the focal distance° of the object-glass by the focal distance of the eye-glass, and the quotient expresses the magnifying power.

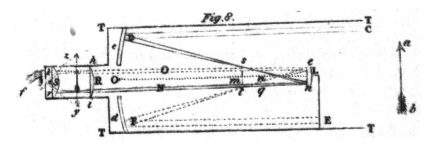

Figure 6.4. The reflecting telescope. (*The Juvenile Encyclopedia*, vol. 3, 3rd plate of "Optics." © The Bodleian Libraries, University of Oxford. (Vet.) 3985 e.19–24.)

An inconvenience was found to attend the use of this instrument, as when any extraordinary magnifying power was wanted the field of view, and even the image, was found to be tinged with different colours. The reason of this will be plain, when I come to treat of the prism and the prismatic colours. You will then see that if a lens is very convex, the edge acts like a prism, and separates the component particles of light, which are differently coloured, and consequently a round circle of different coloured rays is produced. To remedy this, Mr. Dolland,° finding that flint and crown glass had different refracting powers, and that crown glass (the common window glass) dispersed the rays of light less than any other, adapted two convex glasses of crown glass to a double concave of flint glass (which has the greatest dispersive power) so as exactly to fit, and by that means made them counteract each other, so that the field of view is presented perfectly colourless. These telescopes, therefore, are called *achromatic* (or colourless) telescopes.

The reflecting telescope performs, by reflecting the rays issuing from any object, what the last did by refracting them. Let *ab* (see Figure 6.4) be a distant object to be viewed; parallel rays issuing from it, as *ac* and *bd*, will be reflected by the metallic concave mirror *cd* to *st*, and there brought to a focus, with the image a little further and inverted, agreeably to the effect of a concave mirror on light ⌐...⌐. The hole in the mirror *cd* does not distort or hurt the image *st*, it only loses a little light. Nor do the rays stop at the image *st*; they go on, and cross, a little before they reach the small concave mirror *en*. From this mirror the rays are reflected nearly parallel through the hole *O*, in the large mirror, to *R*; there they are met by the plano-convex lens *hi*, which brings them to a convergence at *S*, and paints the image in the small tube of the telescope close to the eye. Having, by this lens, and the two mirrors, brought the image of the object so near, it only remains to magnify this image by the eye-glass *kr*, by which it will appear as large as *zy*.

To produce this effect, it is necessary that the large mirror be ground so as to have its focus a little short of the small mirror, as at *q*, and that the small mirror should be of such concavity as to send the rays a little converging through the hole *o*, that the lens *bi* should be of such convexity as to bring those converging rays to an image at *S*, and that the eye-glass *kr* should be of such a focal length, and so placed in the tube, that its focus may just enter the eye through the small hole in the end of the tube.

To adapt the instrument to near or remote objects, or rather to rays that issue from objects converging, diverging, or parallel, a screw, at the end of a long wire, turns on the outside of the tube, to bring the small mirror nearer to, or farther from, the large mirror, and so as to adjust their focuses according to the nearness or remoteness of the objects. ⌞...⌟

To estimate the magnifying power of the reflecting telescope, multiply the focal distance of the large mirror by the distance of the small mirror from the image *S*; then multiply the focal distance of the small mirror by the focal distance of the eye-glass *kr*; then divide these two products by one another, and the quotient is the magnifying power.

Notes

pencil: a set of lines passing through a single point; so, in this case, the set of light rays emerging from a single point.

focal distance: the distance from the centre of the lens to its focus.

Mr. Dolland: John Dolland (1706–1761), a British optician who was awarded the Royal Society's Copley Medal for his work on achromatic lenses, although he was not the first to make them.

Scales Simple and Diagonal

J. F. Heather, 1888

Originally published in 1849, this treatise promised "to put within the reach of all" a description of the scientific instruments of the day. Many subsequent editions followed, and the book was used in British military and naval schools; the author taught at the Royal Military Academy at Woolwich. By the time of this, the fourteenth edition it even, according to the preface, "formed, by authority, part of a midshipman's kit."

J. F. Heather (d. 1886), *A Treatise on Mathematical Instruments: Their construction, Adjustment, Testing, and Use Concisely Explained. Revised, with Additions By Arthur T. Walmisley, M.I.C.E.* (London: 14th edition, 1888), pp. 9–11, 84–86.

Scales of equal parts are used for measuring straight lines, and laying down distances, each part answering for one foot, one yard, one chain, etc., as may be convenient, and the plan will be larger or smaller as the scale contains a smaller or a greater number of parts in an inch.

Scales of equal parts may be divided into three kinds: simply divided scales, diagonal scales, and vernier scales.

Simply-divided Scales.

Simply-divided scales consist of any extent of equal divisions, which are numbered 1, 2, 3, etc., beginning from the second division on the left hand. The first of these primary divisions is subdivided into ten equal parts, and from these last divisions the scale is named. Thus it is called a scale of 30, when 30 of these small parts are equal to one inch. If, then, these subdivisions be taken as units, each to represent one mile, for instance, or one chain, or one foot, etc., the primary divisions will be so many tens of miles, or of chains, or of feet, etc.; if the subdivisions are taken as tens, the primary divisions will be hundreds; and, if the primary divisions be units, the subdivisions will be tenths.

The accompanying drawing (Figure 6.5) represents six of the simply-divided scales, which are generally placed upon the plain scale. To adapt them to feet and inches, the first primary division is divided duodecimally upon an upper line. To lay down 360, or 36, or 3.6, etc., from any one of these scales, extend the compasses from the primary division numbered 3 to the sixth lower subdivision, reckoning backwards, or towards the left hand. To take off any number of feet and inches, 6 feet 7 inches for instance, extend the compasses from the primary division numbered 6, to the seventh upper subdivision, reckoning backwards, as before.

60	1	2	3	4	5	6	7	8	9	10	1	2	3	4	5	6	7	8	9
50		1	2	3	4	5	6	7	8	9	10	1	2	3	4	5
45		1	2	3	4	5	6	7	8	9	10	1	2	3	4
40			1	2	3	4	5	6	7	8	9	10	1	2
35			1	2	3	4	5	6	7	8	9	10
30			1	2	3	4	5	6	7	8	9

Figure 6.5. Simply-divided scales. (Heather, p. 9. © The Bodleian Libraries, University of Oxford. 1876 f.2.)

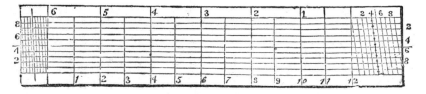

Figure 6.6. Diagonal scales. (Heather, p. 10. © The Bodleian Libraries, University of Oxford. 1876 f.2.)

Diagonal Scales.

In the simply-divided scales one of the primary divisions is subdivided only into ten equal parts, and the parts of any distance which are less than tenths of a primary division cannot be accurately taken off from them; but, by means of a diagonal scale, the parts of any distance which are the hundredths of the primary divisions are correctly indicated, as will easily be understood from its construction, which we proceed to describe.

Draw eleven parallel equidistant lines (see Figure 6.6); divide the upper of these lines into equal parts of the intended length of the primary divisions; and through each of these divisions draw perpendicular lines, cutting all the eleven parallels, and number these primary divisions, 1, 2, 3, etc., beginning from the second.

Subdivide the first of these primary divisions into ten equal parts, both upon the highest and lowest of the eleven parallel lines, and let these subdivisions be reckoned in the opposite direction to the primary divisions, as in the simply-divided scales.

Draw the diagonal lines from the tenth subdivision below to the ninth above, from the ninth below to the eighth above, and so on, till we come to a line from the first below to the zero point above. Then, since these diagonal lines are all parallel, and consequently everywhere equidistant, the distance between any two of them in succession, measured upon any of the eleven parallel lines which they intersect, is the same as this distance measured upon the highest or lowest of these lines, that is, as one of the subdivisions before mentioned. But the distance between the perpendicular, which passes through the zero point, and the diagonal through the same point, being nothing on the highest line, and equal to one of the subdivisions on the lowest line, is equal (Euclid ,book, 6, proposition 4)° to one-tenth of a subdivision on the second line, to two-tenths of a subdivision on the third, and so on; so that this, and consequently each of the other diagonal lines, as it reaches each successive parallel, separates further from the perpendicular through the zero point by one-tenth of the extent of a subdivsion, or one-hundredth of the extent of a primary division. Our figure (6.6) represents the two diagonal scales which are usually placed upon the plain scale of six inches in length. In one, the distances between the primary divisions are each half an inch, and in the other a quarter of an inch. The parallel next to the figures numbering these divisions must be considered the highest or first parallel in each of these scales to accord with the above description.

The primary divisions being taken for units, to set off the number 5.74 by the diagonal scale. Set one foot of the compasses on the point where the fifth parallel cuts the eighth diagonal lines, and extend the other foot to the point where the same parallel cuts the sixth vertical lines.

The primary divisions being reckoned as tens, to take off the number 46.7. Extend the compasses from the point where the eighth parallel cuts the seventh diagonal to the point where it cuts the fifth vertical.

The primary divisions being hundreds, to take off the number 253. Extend the compasses from the point where the fourth parallel cuts the sixth diagonal to the point where it cuts the third vertical.

Now, since the first of the parallels, of the diagonals, and of the verticals indicate the zero points for the third, second, and first figures respectively, the second of each of them stands for, and is marked, 1, the third, 2, and so on, and we have the following.

General Rule.

To take off any number to three places of figures upon a diagonal scale. On the parallel indicated by the third figure, measure from the diagonal indicated by the second figure to the vertical indicated by the first.

Note

Euclid book 6, proposition 4, states that if two triangles have the same set of angles, then their sides are proportional, with corresponding sides being opposite corresponding angles.

Making a Star Clock

Roy Worvill, 1974

Roy Worvill published a number of guides to amateur astronomy in the 1960s and 1970s, including some for young children. Here he gives a wonderfully homely account, complete with cardboard and paper fastener, of the instrument sometimes called a "nocturnal" but here evocatively dubbed the "star clock." It has its roots in the books on "dialing" of the sixteenth and seventeenth centuries, such as Thomas Fale's, above.

Roy Worvill (1914–2003), *Telescope Making for Beginners* (London, 1974, 1976), pp. 70–73.

Making a Star Clock

The sun-dial has the obvious drawback that it is quite useless when the sky is cloudy and at any time when the sun is below the horizon. Indeed, before clocks were invented a cloudy sky made time-keeping a difficult operation since clouds hide the stars as well as the sun. Nevertheless, there are many clear nights, even in the uncertain climate of Britain, and the star clock or nocturnal, to use its old name, then came into its own.

The east-to-west movement of the stars, like that of the sun, is an appearance produced by the earth's daily rotation, and so the stars can be used at night for telling time as the sun can be by day.

In some respects the stars have an advantage, for there are star-groups, or constellations, which never sink below our horizon. To find these we must look towards the northern part of the sky. There, at a height above our horizon equal to our latitude, measured in degrees, we shall find the Pole Star. In fact the Pole Star is not exactly at the north pole in the sky, but it is very near it. The Pole Star appears to remain still, with the other stars circling about it every twenty-four hours as the earth carries us round. Since the Pole Star does not mark the position of the pole exactly it does, in fact, make a very small circle itself. But for our present purpose we can regard it as fixed.

Around the Pole Star there are a number of constellations which never set. These are called the circumpolar groups. The most familiar of them are the seven bright ones which make up the Plough or Dipper, though they are actually part of a much larger constellation called the Great Bear, or Ursa Major, to use its Latin name. The seven stars of the Plough are the hand of our clock and the sky itself is the dial. In particular we use the two stars called the Pointers which lie at the outer edge of the group, the side of the Dipper or saucepan farthest away from the handle. These stars get their name because they point roughly to the Pole Star.

We shall find them rather low down in the northern sky if we look for them in the evening during autumn. In spring, on the other hand, they are to be found high overhead, even if we look for them at the same hour. The reason for this is that the earth travels round the sun a certain distance each day. Measured in miles it is a very long way, for we are moving at just over 18 miles per second, but measured in degrees it is only about one degree daily. The complete circuit of the sun, which takes a year, represents a turn of 360 degrees, and of this a tiny fraction under one degree is covered every day. As a result of these two movements the stars of the Plough make one complete turn round the Pole Star every day as the earth spins on its axis, and a little bit more, approximately one degree, as a result of the other turning movement caused by the earth's revolution round the sun. Because of the little extra movement we find that from night to night, if we look for them at exactly the same time, they have moved a little farther in their circular track round the celestial pole. This complicates our star clock a little, but not too much, since we can make the necessary adjustment for the date. It is, however, rather like having a clock which not only moves its hands but also has a dial which moves slowly round once in a year.

To make the star clock we need the following: a piece of cardboard about sixteen inches long and eight inches wide, a round-headed paper fastener (not a wire clip), a pair of compasses, scissors, a knife and a cutting-board. If you are handy with a fret-saw you can make a better one by using plywood or some other material more rigid than cardboard. Some

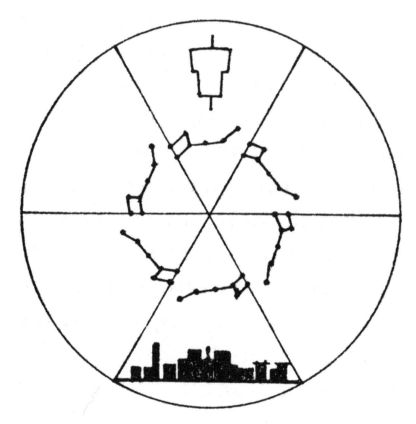

Figure 6.7. The construction of the star clock. (Worvill, p. 72.)

of the ones made centuries ago were of brass and beautifully engraved and decorated. These are still to be found in museums with collections of old scientific instruments.

To make the nocturnal, first cut the sheet of cardboard into two equal squares, each having sides of eight inches. From one of the two squares a circle of about three inches radius is cut out and a small "window" hole, about half an inch square, is cut in the position shown in the diagram (Figure 6.7) between two of the radius lines which mark out angles of sixty degrees. In the sector of the circle which is diametrically opposite the window hole draw in the straight line joining the two ends of the diameters. This line will always indicate the position of your horizon. The centre of the circle shows the position of the celestial pole, marked

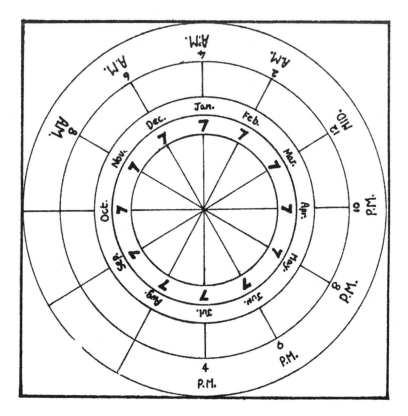

Figure 6.8. The construction of the star clock, continued. (Worvill, p. 73.)

approximately, as we have seen, by the Pole Star. The seven stars of the Dipper are drawn in each of the six sectors of the circle as shown in Figure 6.7, indicating the movement they follow in the course of every twenty-four hours.

On the other cardboard square draw the five concentric circles shown in Figure 6.8, the largest one of radius four inches and the others, proceeding inwards, of three and a half inches, three inches, two and a half inches and two inches. Draw a diameter to the outer circle and mark off angles of thirty degrees so that there are twelve spaces or sectors. On the outer circle these are marked with two-hourly divisions from twelve o'clock midnight throughout the twenty-four hours, although of course the nocturnal can only be used when it is dark enough to see the stars.

Place the small cardboard disc centrally over the large one and make a mark on the lower one through the small window hole. In the ring where the mark appears write the months of the year, the dates showing the seventh day of the month. March 7th should be marked on the line which shows midnight on the hour circle. That is the time when the two stars of the Pointers are in a vertical line with your horizon. You will see from the diagram that the months are marked from March onwards following a clockwise direction while the hours from midnight go anti-clockwise.

The smaller circle is then attached with a paper fastener, through a small central hole to the larger one so that it will revolve easily, but not too loosely, over the lower dial. This completes the making of the star clock.

To use it, you hold it in front of you so that the centre of the dial, marked by the fastener, is approximately over the Pole Star. The smaller dial is turned so that it shows the date, as nearly as possible through the small window hole. The whole clock is then turned so that the horizon line is level at the lowest position of the circle. Look at the Dipper stars to see which of the six star diagrams corresponds most accurately to the position of the Dipper. A line from the Pointers in that section of the star diagram is followed to the outside edge of the dial, and this will show you the time. By this means it should be possible to tell the time to within about ten or fifteen minutes. Of course your watch will do this more accurately, but even the best watches can run down or go wrong, and it is interesting to remember that the first clocks and watches were so unreliable that a small pocket sun-dial or nocturnal was often carried in those days to check the time-keeping of the watch or clock! Apart from any practical use as a time-keeper the nocturnal, like the sun-dial, will give you a better understanding of the changing appearance of the sky through the seasons.

PC Astronomy

Peter Duffet-Smith, 1997

A follow-up to earlier books by the same author—*Astronomy with Your Pocket Calculator* and *Astronomy with your Personal Computer*—this volume, *Easy PC Astronomy*, was geared toward the efficient and convenient making of sometimes complex astronomical calculations rather than toward producing glossy displays or pictures. It discussed a wide variety of often quite technical computations connected with the practice of amateur astronomy and continued the practical theme we have seen in the making of sundials and star clocks.

Peter Duffett-Smith (dates unknown), *Easy PC Astronomy* (Cambridge, 1997), pp. 48–51. ⓒ Cambridge University Press 1997; reproduced with permission.

Precession

Anyone who has played with a spinning top as a child will know that its axis slowly gyrates about the vertical direction as it spins. This phenomenon is common to any spinning object which is under the influence of an external couple, and is called *gyroscopic precession*. In the case of the spinning top, the couple is provided by the weight of the top which pulls vertically downwards through its centre of gravity. The Earth is also a spinning object and it, too, exhibits the same precessional behaviour. The gravitational attractions of the Moon and the Sun, acting on the equatorial bulges of the slightly non-spherical Earth, cause the couple which results in the north–south axis precessing slowly about a line through the centre of the Earth and perpendicular to the plane of the ecliptic° with a period of about 26,000 years. The effect of this *luni-solar precession* is to make the direction of the line of intersection of the rising node of the celestial equator on the plane of the ecliptic (the first point of Aries of vernal equinox)° move steadily at a rate of about 50 arcseconds per year.

There is also another, smaller, precession from the gravitational influences of the planets. If the celestial equator were fixed (i.e. not perturbed by luni-solar precession), their combined effect would be to cause the equinox to move by about 12 arcseconds per century, and to decrease the obliquity of the ecliptic° by about 47 arcseconds per century. This is called *planetary precession*. Both luni-solar and planetary precession are usually calculated together as *general precession*.

The positions of "fixed" stars and other celestial objects are often defined by their right ascensions and declinations, or by their ecliptic longitudes and latitudes. In either case, the reference direction is that of the first point of Aries, so general precession causes these coordinates to change slowly with time. Hence it is necessary to specify the moment, or *epoch*, at which a particular pair of coordinates is valid. A correction can then be applied to convert the coordinates to values which are valid at another epoch. AstroScript incorporates an algorithm for performing this correction for general precession rigorously ₍...₎; it converts the currently held right ascension and declination into their new values. It asks you to supply the two precessional epochs if you have not already done so.

The AstroScript algorithm for general precession follows that laid out in the *Explanatory Supplement to the Astronomical Almanac* ₍...₎. If (α_1, δ_1) are the right ascension and declination valid at epoch ϵ_1, the column vector $\mathbf{r}_1$ is first obtained by

$$\mathbf{r}_1 = \begin{pmatrix} \cos\alpha_1 \cos\delta_1 \\ \sin\alpha_1 \cos\delta_1 \\ \sin\delta_1 \end{pmatrix}.$$

This is then multiplied by the precession matrix, $\mathbf{P}$, to form the new column vector, $\mathbf{r}_2$, appropriate for the new epoch ϵ_2 as follows:

$$\mathbf{r}_2 = \mathbf{P} \cdot \mathbf{r}_1.$$

If the components of $\mathbf{r}_2$ are

$$\mathbf{r}_2 = \begin{pmatrix} m \\ n \\ p \end{pmatrix},$$

the new coordinates, (α_2, δ_2), are obtained from $\mathbf{r}_2$ by the equations

$$\tan\alpha_2 = n/m, \text{ and } \sin\delta_2 = p.$$

The precession matrix $\mathbf{P}$ is given by

$$\mathbf{P} = \begin{pmatrix} cx \cdot ct \cdot cz - sx \cdot sz & -sx \cdot ct \cdot cz - cx \cdot sz & -st \cdot cz \\ cx \cdot ct \cdot sz + sx \cdot cz & -sx \cdot ct \cdot sz + cx \cdot cz & -st \cdot sz \\ cx \cdot st & -sx \cdot st & ct \end{pmatrix},$$

where $cx = \cos\zeta_A$, $sx = \sin\zeta_A$, $cz = \cos z_A$, $sz = \sin z_A$, $ct = \cos\theta_A$, and $st = \sin\theta_A$. These arguments are calculated from the following expressions involving the interval in Julian centuries of 36525 days between ϵ_0, the fundamental epoch J2000.0 (1.5 Jan. 2000), and the two epochs ϵ_1 and ϵ_2, where

$$T = [\mathrm{JD}(\epsilon_1) - \mathrm{JD}(\epsilon_0)]/36525,$$

$$t = [\mathrm{JD}(\epsilon_2) - \mathrm{JD}(\epsilon_1)]/36525,$$

JD(ϵ) represents the Julian day number of epoch ϵ, and JD(ϵ_0) = 2 451 545.0,

$$\zeta_A = (2306.2181 + 1.39656T - 0.000139T^2)t$$

$$+(0.30188 - 0.000344T)t^2 + 0.017998t^3$$

$$z_A = (2306.2181 + 1.39656T - 0.000139T^2)t$$

$$+(1.09468 + 0.000066T)t^2 + 0.018203t^3$$

$$\theta_A = (2004.3109 - 0.85330T - 0.000217T^2)t$$

$$+(-.42665 - 0.000217T)t^2 - 0.041833t^3.$$

ζ_A, z_A, and θ_A are all given in arcseconds by the above expressions.

Note

ecliptic: the plane of the earth's orbit around the sun.

the line of intersection ...: Another way to describe this is as the line in which the plane of the earth's orbit and the plane of the equator intersect.

the obliquity of the ecliptic: the angle between the plane of the earth's equator and that of its orbit.

❧ 7 ❧

"How Fine a Mind": Mathematicians Past

INTEREST IN MATHEMATICIANS PAST BEGAN IN THE SIXTEENTH AND SEVEN-
teenth centuries with the rediscovery, publication, and commentary on
ancient mathematical texts, part of the more general Renaissance concern
to rediscover and surpass ancient learning (and a process which continues
to this day: "new" texts by Archimedes have been in the news in the last few
years). This side of mathematical history writing is represented elsewhere
in this book by the passage from Euclid in Chapter 4.

Also on display in other chapters is guessing—or, to be kinder,
speculation—about the origins of mathematical techniques. We've seen
both Joseph Fenn (Chapter 2) and Edmund Scarburgh (Chapter 4) try their
hand at that in the contexts of algebra and geometry, and in this chapter
we'll see Charles Hutton do something similar for the supposed beginnings
of arithmetic.

Both of these strategies receive their modern transformation in such a
work as Thomas Heath's edition of Euclid, where detailed research into ancient mathematical techniques and discoveries complements precise work
on the ancient texts themselves. This textual and thematic approach to the
mathematical past has been both fruitful and important and promises to
continue in the future.

In popular writing it is complemented by an equally common—perhaps
a more common—kind of mathematical history: biography. Several kinds
of biography are showcased here: celebratory accounts of great men (Newton, in this case), eulogy, and the drier approaches of writers of historical
textbooks, who until well into the twentieth century still relied very much
on (sometimes dubious) anecdotes as a way to depict the characters and
the activities of the past. This approach, again, blossoms into the modern
school of mathematical biography—like Robert Kanigel's—in which the
individual scene may be a springboard to a really meaningful exploration
of themes or technical content in the mathematics of the past.

In this chapter, too, we witness the abiding tendency to "domesticate" the mathematics of (for the historians) distant times and places by finding modern or western equivalents for it, but also the slow rise of a consciousness that the long-excluded—women, the poorly educated, the subjects of colonial rule—had a place in history as creators of mathematics.

The Labyrinth and Abyss of Infinity

Voltaire, 1733

Voltaire's 1733 *Letters Concerning the English Nation* (published in French a year later as *Lettres philosophiques*) are sometimes called the first work of the Enlightenment; they commented on many aspects of English life, from science and religion to literature and theatre. Three of the twenty-four letters discussed Sir Isaac Newton at length; Voltaire was one of his greatest admirers—or propagandists— and was famously impressed by the fact that Newton was buried "like a king" at a state funeral. The Newtonian subjects covered in the *Letters* included optics, chronology, and anti-Cartesianism; here we see Voltaire describing the genesis of integral and differential calculus with an eye to the history of the subject and the subsequent dispute rather than its technical details. (For another popular account of matters Newtonian from the same decade, see the extract from Algarotti in Chapter 5.)

Voltaire (1694–1778), trans. John Lockman (1698–1771), *Letters concerning the English nation* (London, 1733), pp. 151–155 (letter 17).

The Labyrinth and Abyss of Infinity, is also a new Course Sir Isaac Newton has gone through, and we are obliged to him for the Clue by whose Assistance we are enabled to trace its various Windings.

Descartes got the Start of him also in this astonishing Invention. He advanced with mighty Steps in his Geometry, and was arrived at the very Borders of Infinity, but went no farther. Dr. Wallis, about the Middle of the last Century, was the first who reduced a Fraction by a perpetual Division to an infinite Series.

The Lord Brounker employed this Series to square° the Hyperbola.

Mercator° published a Demonstration of this Quadrature,° much about which Time, Sir Isaac Newton, being then twenty three Years of Age, had invented a general Method to perform, on all geometrical Curves, what had just before been tried on the Hyperbola.

'Tis to this Method of subjecting, everywhere, Infinity to algebraical Calculations, that the Name is given of differential Calculations or of Fluxions, and integral Calculation. 'Tis the Art of numbering and measuring exactly a Thing whose Existence cannot be conceived.

And, indeed, would you not imagine that a Man laughed at you, who should declare that there are Lines infinitely great which form an Angle infinitely little?

That a ˌstraightˌ Line, which is a ˌstraightˌ Line so long as it is finite, by changing infinitely little its Direction, becomes an infinite Curve; and that a Curve may become infinitely less than another Curve?

That there are infinite Squares, infinite Cubes; and Infinities of Infinities all greater than one another, and the last but one of which, is nothing in Comparison of the last?

All these Things which at first appear to be the utmost Excess of Frenzy, are in reality an Effort of the Subtlety and Extent of the human Mind, and the Art of finding Truths which till then had been unknown.

This so bold Edifice is even founded on simple Ideas. The Business is to measure the Diagonal of a Square, to give the Area of a Curve, to find the square Root of a Number, which has none in common Arithmetic. After all, the Imagination ought not to be startled any more at so many Orders of Infinites, than at the so well known Proposition, *viz.* that Curve Lines may always be made to pass between a Circle and a Tangent; or at that other, namely that Matter is divisible *in infinitum.* These two Truths have been demonstrated many Years, and are no less incomprehensible than the Things we have been speaking of.

For many Years the Invention of this famous Calculation was denied Sir Isaac Newton. In Germany Mr. Leibniz° was considered as the Inventor of the Differences or Moments, called Fluxions, and Mr. Bernoulli° claim'd the integral Calculation. However, Sir Isaac is now thought to have first made the Discovery, and the other two have the Glory of having once made the World doubt whether 'twas to be ascribed to him or them.

◦

Be this as it will, 'tis by the Help of this Geometry of Infinities that Sir Isaac Newton attained to the most sublime Discoveries.

Notes

square: find the area under. William Brouncker (1620–1684) was the first president of the
    Royal Society.

Mercator: not the map maker but Nicolaus Mercator (1620?–1687), a Danish mathematician
    who spent much of his life in England after leaving Copenhagen to escape the plague.

Quadrature: area finding.

Mr. Leibniz: Gottfried Wilhelm von Leibniz (1646–1716), the German philosopher and
    mathematician.

Mr. Bernoulli: the Swiss mathematician Jacob Bernoulli (1654–1705).

"It Must Have Commenced with Mankind"

Charles Hutton, 1796

No selection could do justice to the magnificent *Dictionary* compiled by the prolific
English mathematician and mathematical writer Charles Hutton, which contains
some of the first really accessible mathematical history to be written in English.
The section chosen here illustrates a thematic rather than a biographical approach
to mathematical history and gives a hint of the range of Hutton's learning. It may
be compared with the other speculations in this volume, by Joseph Fenn and by
Edmund Scarburgh, on the origins of algebra and geometry.

Charles Hutton (1737–1823), *A Mathematical and Philosophical Dictionary: Containing an
Explanation of the Terms, and an Account of the Several Subjects, comprised under the heads
Mathematics, Astronomy, and Philosophy both natural and experimental: with an Historical
Account of the Rise, Progress, and Present State of these Sciences: also Memoirs of the Lives and
Writings of the Most Eminent Authors, both ancient and modern, who by their discoveries
or improvements have contributed to the advancement of them* (London, 1796), vol. 1,
pp. 142–143.

Arithmetic: the art and science of numbers; or, that part of mathematics
which considers their powers and properties, and teaches how to compute
or calculate truly, and with ease and expedition. It is by some authors also
defined the science of discrete quantity. Arithmetic consists chiefly in the
four principal rules or operations of Addition, Subtraction, Multiplication,
and Division; to which may perhaps be added involution and evolution,
or raising of powers and extraction of roots. But besides these, for the
facilitating and expediting of computations, mercantile, astronomical, etc.,
many other useful rules have been contrived, which are applications

of the former, such as, the rules of proportion, progression, alligation, false position, fellowship, interest, barter, rebate, equation of payments, reduction, tare and tret, etc. Besides the doctrine of the curious and abstract properties of numbers.

Very little is known of the origin and invention of arithmetic. In fact it must have commenced with mankind, or as soon as they began to hold any sort of commerce together; and must have undergone continual improvements, as occasion was given by the extension of commerce, and by the discovery and cultivation of other sciences. It is therefore very probable that the art has been greatly indebted to the Phœnicians or Tyrians; and indeed Proclus, in his commentary on the first book of Euclid, says, that the Phœnicians, by reason of their traffic and commerce, were accounted the first inventors of Arithmetic. From Asia the art passed into Egypt, whither it was carried by Abraham, according to the opinion of Josephus. Here it was greatly cultivated and improved; insomuch that a considerable part of the Egyptian philosophy and theology seems to have turned altogether upon numbers. Hence those wonders related by them about unity, trinity, with the numbers 4, 7, 9, etc. In effect, Kircher, in his *Oedipus Ægyptus* shews, that the Egyptians explained every thing by numbers; Pythagoras himself affirming, that the nature of numbers pervades the whole universe; and that the knowledge of numbers is the knowledge of the deity.

From Egypt arithmetic was transmitted to the Greeks, by means of Pythagoras and other travellers; amongst whom it was greatly cultivated and improved, as appears by the writings of Euclid, Archimedes, and others: with these improvements it passed to the Romans, and from them it has descended to us.

The nature of the arithmetic however that is now in use, is very different from that above alluded to; this art having undergone a total alteration by the introduction of the Arabic notation, about 800 years since, into Europe: so that nothing now remains of use from the Greeks, but the theory and abstract properties of numbers, which have no dependence on the peculiar nature of any particular scale or mode of notation. That used by the Hebrews, Greeks, and Romans, was chiefly by means of the letters of their alphabets. The Greeks, particularly, had two different methods; the first of these was much the same with the Roman notation, which is sufficiently well known, being still in common use with us, to denote dates, chapters and sections of books, etc. Afterwards they had a better method, in which the first nine letters of their alphabet represented the first numbers, from one to nine, and the next nine letters represented any number of tens, from one to nine, that is, 10, 20, 30, etc., to 90. Any number of hundreds they expressed by other letters, supplying what they wanted with some other marks or characters: and in this order they went on, using the same letters again, with some different marks, to express thousands, tens of thousands, hundreds of thousands, etc: In which it is evident that they approached very near to the more perfect decuple scale of progression used by the Arabians, and who acknowledge that they had received it from the Indians. Archimedes also invented another peculiar scale and notation of his own, which he employed in his Arenarius, to compute the number of the sands. In the 2nd century of Christianity lived Claudius Ptolemy, who, it is supposed, invented the sexagesimal division of numbers, with its peculiar notation and operations: a mode of computation still used in astronomy etc, for the subdivisions of the degrees of circles. Those notations however were ill adapted to the practical operations of arithmetic: and hence it is that the art advanced but very little in this part; for, setting aside Euclid, who has given many plain and useful properties of numbers in his Elements, and Archimedes, in his Arenarius, they mostly consist in dry and tedious distinctions and divisions of numbers; as appears from the treatises of Nicomachus, supposed to be written in the 3rd century of Rome, and published at Paris in 1538; as also that of Boethius, written at Rome in the 6th century of Christ. A compendium of the ancient arithmetic, written in Greek, by Psellus, in the 9th century, was published in Latin by Xylander, in 1556. A similar work was written soon after in Greek by Jodocus Willichius; and a more ample work of the same kind was written by Jordanus, in the year 1200, and published with a comment by Faber Stapulensis in 1480.

Since the introduction of the Indian notation into Europe, about the 10th century, arithmetic has greatly changed its form, the whole algorithm, or practical operations with numbers, being quite altered, as the notation required; and the authors of arithmetic have gradually become more and more numerous. This method was brought into Spain by the Moors or Saracens; whither the learned men from all parts of Europe repaired, to learn the arts and sciences of them. This, Dr. Wallis proves, began about the year 1000; particularly that a monk, called Gilbert, afterwards pope, by the name of Sylvester II, who died in the year 1003, brought this art from Spain into France, long before the date of his death: and that it was known in Britain before the year 1150, where it was brought into common use before 1250, as appears by the treatise of arithmetic of Johannes de Sacro Bosco, or Halifax, who died about 1256. Since that time, the principal writers on this art have been, Barlaam, Lucas de Burgo, Tonstall, Aventinus, Purbach, Cardan, Scheubelius, Tartalia, Faber, Stifelius, Recorde, Ramus, Maurolycus, Hemischius, Peletarius, Stevinus, Xylander, Kersey, Snellius, Tacquet, Clavius, Metius, Gemma Frisius, Buteo, Ursinus, Romanus, Napier, Ceulen, Wingate, Kepler, Briggs, Ulacq, Oughtred, Cruger, Van Schooten, Wallis, Dee, Newton, Morland, Moore, Jeake, Ward, Hatton, Malcolm, etc, etc.

Kepler's Astronomical Publications

Robert Small, 1804

Robert Small, Doctor of Divinity and Fellow of the Royal Society of Edinburgh (indeed, one of that society's founding fellows in 1783), contributed to the *Transactions* of the society on mathematical subjects. His other publications were a work of political statistics and a sermon connected with the charitable support of Sunday schools; in 1791 he was Moderator of the General Assembly of the Church of Scotland. His interest in the history of mathematics was at the time a relatively uncommon way of representing mathematics to nonspecialist readers, but it was a sign of things to come.

Robert Small (1732–1808), *An Account of the Astronomical Discoveries of Kepler: Including An Historical Review of the Systems which had succesively prevailed before his time* (London, 1804), pp. 146–159.

The first fruits of ⌊Kepler's⌋ application to astronomical studies appeared in his *Mysterium Cosmographicum*, published ⌊in 1596,⌋ about two years after his settlement in Graz. Though he had adopted the Copernican arrangement° of the planets, the principles of connection subsisting between its parts appeared to him to be in a great degree unknown. No causes especially were assigned for the different positions of the centre of uniform motion in the different orbits, nor for the irregular distances of the planets from the sun, and no proportion was discovered between these distances and the times of the planetary revolutions. He considered the arrangement as deficient and unsatisfactory, till all those causes should be discovered; and in the work now mentioned he supposed that he had discovered them, at least he traced out several similar analogies, in the Pythagorean, or Platonic doctrines concerning numbers, in the proportions of the regular solids in geometry, and in the divisions of the musical scale; and these analogies seemed to assign the reason why the primary planets should be only six in number.

Hasty and juvenile as this production was, it displayed so many marks of genius, and such indefatigable patience in the toil of calculation, that, on presenting it to T. Brahé,° it procured him the esteem of this illustrious astronomer, and even excited his anxiety for the proper direction of talents so uncommon. Accordingly, not contented with exhorting Kepler to prefer the road of observation to the more uncertain one of theory, T. Brahé added a generous and unsolicited invitation to live with him at Uraniburg, where his whole observations would be open to Kepler's perusal, and those advantages found for making others, which his situation at Graz denied. The opinion of other astronomers concerning this production was no less favourable, and, were there no other evidence of his just and general conceptions, the remark, which afterward led to such important consequences, that the cause of the equant,° whatever it might be, ought to operate universally, is a sufficient attestation of them, and a proof that, even in this early period of his studies, the possibility at least of deducing the equations of a planet from the relation between its different degrees of velocity and distances from the sun had presented itself to his thoughts.

Notwithstanding the ardent desire of Kepler to be admitted to the perusal of T. Brahé's observations, it was probable that the distance of the place, and the difficulty of the journey, to one in Kepler's situation, would have for ever prevented the gratification of it. But the persecutions now arose which drove T. Brahé from his native country, and from which he at last found a refuge in Prague, the capital of Bohemia. Thither, accordingly, Kepler repaired in the year 1600; and, that he might be under no necessity of returning to Graz, he obtained, by T. Brahé's interest with the Emperor, his patron, the appointment of imperial mathematician.

On his arrival at Prague another circumstance occurred, equally important to his success and fame, and which he piously ascribes to the kind
direction of Providence. T. Brahé, with his assistant Longomontanus,° was
employed about his theory of Mars, no doubt induced by the favourable
opportunity of verifying it by observations of this planet in its approaching
opposition in the sign of Leo. Kepler's attention was therefore, of course,
directed to the same planet, where the greater degree of eccentricity
renders its inequalities peculiarly remarkable, and leads with proportional
advantage to the discovery of their laws; and it would have been directed
to observations much less instructive, had those astronomers been engaged
about any other planet.

The former planetary theories were in general unlike, and founded upon
different principles. The ancients, as Ptolemy and his followers, considered
every planet separately, and supposed that their several motions and
inequalities arose from causes peculiar to every orbit, and with which the
other orbits had no connection. The moderns, again, considered the orbits
as connected by a common principle, and, remarking all that was similar
in their motions, endeavoured to derive it from a common cause. But they
disagreed about this principle, for the cause of the second inequalities°
of the planets, according to Copernicus, was the annual revolution of the
earth, while T. Brahé and his disciples ascribed them to the attendance of
the planets upon the sun, in his annual revolution. But, notwithstanding
these and other less important differences, the effects of all the theories, in
representing the motions of the planets, and on the calculations of their
places, were very nearly equal, and Kepler found them almost equally
erroneous: for the longitudes of Mars, calculated for August 1598, and
August 1608, even from the Prutenic tables of Reinholdus,° fell short of
the observed longitudes, the first nearly 4°, and the last no less than nearly
5°; and errors of this magnitude were not peculiar to the Prutenic tables,
for similar and even greater errors were found in all the others.

Notes

Copernican arrangement: with the sun at the centre rather than the Earth.

T. Brahé: the Danish astronomer Tycho Brahe (1546–1601) on whose observational data
    Kepler eventually relied.

equant: a point from which a particular planet appears to move at constant angular speed, or
    a circle about this point.

Longomontanus: Christen Sørensen Longomontanus (1562–1647), Danish astronomer, important for promoting Tycho's theory of planetary motion.

second inequalities: nonuniformities in planetary motion; specifically, those not explained at the first stage of constructing a model of planetary motion.

the Prutenic tables of Reinholdus: the German astronomer and mathematician Erasmus Reinhold (1511–1553) published his astronomical tables, based on Copernicus's theory of planetary motion, in 1551.

Isaac Newton, a Good and Great Man

Anonymous, 1860

The nineteenth century saw the considerable development of the "Newton myth," the publication of more of Newton's papers and the appearance of the first biography based on them. This charming (though not exactly accurate) portrait gives a sense of how Newton was seen at a more popular level; see Figure 7.1 for the accompanying illustration, complete with the famous apple. (George Bernard Shaw, in the following extract, will show us in some ways a fairly similar Newton.) He is, aside from Christopher Wren, the only mathematician among the fifty "greats" in the book.

The Children's Picture-Book of Good and Great Men (London, 1860), pp. 150–156.

Sir Isaac Newton, the greatest philosopher and astronomer of modern days, was born on Christmas day, 1642, at Woolsthorpe, in Lincolnshire.

He was such a very little creature that his mother declared he might have been put into a quart-mug, and so feeble that no one thought he could live. But that poor little weak baby was to be one of our greatest instructors, and to find out more of God's wonderful power and wisdom, in the creation of the earth, and those heavenly bodies that shine out to us in the sky, than anyone had ever before done.

Isaac's mother, who was a widow, carefully nursed him up till he was three years old, when his grandmother took charge of the fatherless child, and after sending him for a while to a day-school in the neighbourhood, placed him, that he might have better teaching, at the Grammar-school of Grantham.

Figure 7.1. Newton meditating on the fall of the apple. (*Picture-book*, p. 155. © The Bodleian Libraries, University of Oxford. Nuneham 2526 e.4.)

He was here chiefly noticed as an ingenious lad, being fond, as boys generally are, of using carpenter's tools, only that he managed them better than boys usually do. Among the various articles that his nimble fingers and thoughtful little head contrived was a water-clock. A box filled with water, which was permitted to escape drop by drop from the lower part, formed the body of this clock, while a piece of wood floating on the surface of the water was so contrived as, by its gradual sinking, to point out, one after the other, the hours of the day, which were marked on another portion of his ingenious contrivance.

At the age of fourteen, he was taken from school to be made a farmer of. But it was soon plain that Isaac was no farmer, nor man of business either. Each market-day that he was sent to Grantham, instead of busying himself to make good bargains for his corn or hay, he slunk off to an attic in the house where he had lodged while at school, and there sat poring over some old mathematical books, till the servant, who had been sent with him, called to take him home. Sometimes he never reached Grantham at all, but would stop by the road-side, quite taken up with a water-mill, or some such machine, till the returning waggon picked him up again. This would never do, and his mother, knowing that his inattention to business proceeded neither from idleness nor perverseness, but from his whole soul being taken up with study, instead of forcing him to be a farmer, sent him to college, where he might have his fill of learning. Trinity College, Cambridge, has the honour of having given young Newton (he was only eighteen) his first training in science.

He was in his right place there, and speedily showed how fine a mind he possessed. Within six years, that is, by the time he had reached his twenty-fifth year, he had made some of his greatest discoveries, including the one for which he is most celebrated—that of finding out how it is that the moon and stars keep their places, circling in wide space around the sun. This had long puzzled learned men. Newton, sitting in his garden one day, saw an apple fall from the tree, and, strangely enough, as he sat there thinking why it should come to the ground, he found out that the very same thing that made it do so was that which worked all the wonders of the regular movements of the heavenly bodies. To understand this would require more learning than you children have got. But people with sufficient learning see very well how it is, and that it is perfectly true.

Newton made many important discoveries in other sciences. But his vast knowledge did not make him think highly of himself. Much as he had discovered of God's works, he knew there was so much more still unknown, that, at the close of his long life he said that to himself he appeared like a child picking up a few shells on the sea-shore, while the "great ocean of truth" lay all undiscovered before him.

In the year 1705 he received the honour of knighthood from Queen Anne.

This eminent philosopher died on the 20th of March, 1727, in the eighty-fifth year of his age, and was buried in Westminster Abbey; his pall was borne by the lord chancellor, two dukes, and three earls.

He was a man of great amiability and goodness. He was a sincere Christian, and spent much of his time in the study of the Holy Scriptures; nor could anything cause him greater grief than to hear the subject of religion spoken of in a light and irreverent manner.

Pythagoras and His Theorem

Thomas L. Heath, 1908

Still the definitive English version of Euclid a century after publication, Thomas Heath's magnificent edition is made with rather different intentions from many earlier versions, and indeed it comes with a wealth and depth of commentary which seem intended to place it outside the category of "popular mathematics." The commentary on Pythagoras's theorem continues for another seventeen pages after the extract given here, which is presented with some simplifications. But Heath made clear that he nonetheless considered the *Elements* above all an *elementary* textbook, indeed the supreme example of its kind, "the greatest . . . that the world is privileged to possess."

Sir Thomas L. Heath (1861–1940), *The Thirteen Books of Euclid's Elements, Translated from the text of Heiberg with introduction and commentary by Sir Thomas L. Heath* (Cambridge, 1908, 1925), vol. 1, pp. 349–352.

"If we listen," says Proclus,° "to those who wish to recount ancient history, we may find some of them referring this theorem to Pythagoras and saying that he sacrificed an ox in honour of his discovery. But for my part, while I admire those who first observed the truth of this theorem, I marvel more at the writer of the Elements, not only because he made it fast by a most lucid demonstration, but because he compelled assent to the still more general theorem by the irrefragable arguments of science in the sixth Book. For in that Book he proves generally that, in right-angled triangles, the figure on the side subtending the right angle is equal to the similar and similarly situated figures described on the sides about the right angle."

In addition, Plutarch, Diogenes Laertius and Athenaeus° agree in attributing this proposition to Pythagoras. It is easy to point out, as does G. Junge, that these are late witnesses, and that the Greek literature which we possess belonging to the first five centuries after Pythagoras contains no statement specifying this or any other particular great geometrical discovery as due to him. Yet the distich of Apollodorus the "calculator," whose date (though it cannot be fixed) is at least earlier than that of Plutarch and presumably of Cicero,° is quite definite as to the existence of *one* "famous proposition" discovered by Pythagoras, whatever it was. Nor does Cicero, in commenting apparently on the verses, seem to dispute the fact of the geometrical discovery, but only the story of the sacrifice. Junge naturally emphasises the apparent uncertainty in the statements of Plutarch and Proclus. But, as I read the passages of Plutarch, I see nothing in them inconsistent with the supposition that Plutarch unhesitatingly accepted as discoveries of Pythagoras *both* the theorem of the square of the hypotenuse and the problem of the application of an area, and the only doubt he felt was as to which of the two discoveries was the more appropriate occasion for the supposed sacrifice.

There is also other evidence not without bearing on the question. The theorem is closely connected with the whole of the matter of Euclid Book II, in which one of the most prominent features is the use of the *gnomon.*° Now the gnomon was a well-understood term with the Pythagoreans. Aristotle also clearly attributes to the Pythagoreans the placing of odd numbers as *gnomons* round successive squares beginning with 1, thereby forming new squares, while in another place the word *gnomon* occurs in the same (obviously familiar) sense: "e.g. a square, when a gnomon is placed round it, is increased in size but is not altered in form." The inference must therefore be that practically the whole doctrine of Book II is Pythagorean.

Again Heron (?3rd cent. A.D.), like Proclus, credits Pythagoras with a general rule for forming right-angled triangles with rational whole numbers for sides. Lastly, the "summary" of Proclus appears to credit Pythagoras with the discover of the theory, or study, of irrationals. But it is now more or less agreed that the reading here should be ₍emended:₎ "of proportionals," and that the author intended to attribute to Pythagoras a theory of *proportion*, i.e. the (arithmetical) theory of proportion applicable only to commensurable magnitudes, as distinct from the theory of Euclid Book V, which was due to Eudoxus.° It is not however disputed that the *Pythagoreans* discovered the irrational.

Now everything goes to show that this discovery of the irrational was made with reference to $\sqrt{2}$, the ratio of the diagonal of a square to its side. It is clear that this presupposes the knowledge that I.47 is true of an isosceles right-angled tringle; and the fact that some triangles of which it had been discovered to be true were *rational* right-angled triangles was doubtless what suggested the inquiry whether the ratio between the lengths of the diagonal and the side of a square could also be expressed in whole numbers. On the whole, therefore, I see no sufficient reason to question the tradition that, *so far as Greek geometry is concerned* (the possibly priority of the discovery of the same proposition in India will be considered later), Pythagoras was the first to introduce the theorem of I.47 and to give a general proof of it.

Notes

Proclus: Proclus Diadochus (411–485), a Greek philosopher important for his commentaries on various mathematicians.

Plutarch, Diogenes Laertius and Athenaeus: Greek writers on philosophy and (Athenaeus) rhetoric and grammar of, respectively, the 1st–2nd, 3rd, and 2nd–3rd centuries.

Cicero: the great Roman orator and statesman (106–43 BC).

gnomon: "the part of a parallelogram which remains after a similar parallelogram is taken away from one of its corners." (*OED*).

Eudoxus of Cnidus: Greek mathematician and astronomer (408–355 BC), to whom (part of) the theory of ratios in Euclid's *Elements* is often attributed.

Seki Kōwa

David Eugene Smith and Yoshio Mikami, 1914

David Eugene Smith was a professor of mathematics at Teachers College, Columbia University, involved with the American Mathematical Society, the International Commission on the Teaching of Mathematics, the Mathematical Association of America, and the History of Science Society. He took a keen interest in the history of mathematics, publishing a sourcebook and a textbook on the subject. Yoshio Mikami published a history of mathematics in China and Japan.

In our extract the authors give an outline of the life and achievements of Seki Kōwa, one of the great Japanese mathematicians of the seventeenth century, with an example of his work. Although the approach is basically biographical, there is much more mathematical detail here than in earlier works like Small's discussion of Kepler; indeed, the presentation perhaps has more in common with Thomas Heath's annotated translation of Euclid. The footnotes in this extract are by the original authors.

David Eugene Smith (1860–1944) and Yoshio Mikami (1875–1950), *A History of Japanese Mathematics* (Chicago, 1914), pp. 91–100.

In the third month according to the lunar calendar, in the year 1642 of our era, a son was born to Uchiyama Shichibei, a member of the *samurai* class living at Fujioka in the province of Kōzuke. While still in his infancy this child, a younger son of his parents, was adopted into another noble family, that of Seki Gorozayemon, and hence there was given to him the name of Seki by which he is commonly known to the world. Seki Shinsuke Kōwa was born in the same year in which Galileo died, and at a time of great activity in the mathematical world both of the East and the West. And just as Newton, in considering the labors of such of his immediate predecessors as Kepler, Cavalieri, Descartes, Fermat, and Barrow, was able to say that he had stood upon the shoulders of giants, so Seki came at an auspicious time for a great mathematical advance in Japan, with the labors of Yoshida, Imamura, Isomura, Muramatsu, and Sawaguchi° upon which to build. The coincidence of birth seems all the more significant because of the possible similarity of achievement, Newton having invented the calculus of fluxions in the West, while Seki possibly invented the *yenri* or "circle principle" in the East, each designed to accomplish much the same purpose, and each destined to material improvement in later generations. The *yenri* is not any too well known and it is somewhat difficult to judge of its comparative value, Japanese scholars themselves being undecided as to the relative merits of this form of the calculus and that given to the world by Newton and Leibniz.

Seki's great abilities showed themselves at an early age. The story goes that when he was only five he pointed out the errors of his elders in certain calculations which were being discussed in his presence, and that the people so marveled at his attainments that they gave him the title of divine child.

Another story relates that when he was but nine years of age, Seki one time saw a servant studying the *Jinkō-ki* of Yoshida. And when the servant was perplexed over a certain problem, Seki volunteered to help him, and easily showed him the proper solution. This second story varies with the narrator, Kamizawa Teikan telling us that the servant first interested the youthful Seki in the arithmetic of the *Jinkō-ki*, and then taught him his first mathematics. Others say that Seki learned mathematics from the great teacher Takahara Kisshu, who ⌊...⌋ had sat at the feet of Mōri as one of his *san-shi*, although this belief is not generally held. Most writers agree that he was self-made and self-educated, his works showing no apparent influence of other teachers, but on the contrary displaying an originality that may well have led him to instruct himself from his youth up. Whatever may have been his early training Seki must have progressed very rapidly, for he early acquired a library of the standard Japanese and Chinese works on mathematics, and learned, apparently from the *Suan-hsiao Chi-mêng*, the method of solving the numerical higher equation. And with this progress in learning came a popular appreciation that soon surrounded him with pupils and that gave to him the title of The Arithmetical Sage. In due time he, as a descendent of the *samurai* class, served in public capacity, his office being that of examiner of accounts to the Lord of Kōshū, just as Newton became master of the Mint under Queen Anne. When his lord became heir to the Shōgun, Seki became a Shogunate *samurai*, and in 1704 was given a position of honor as master of ceremonies in the Shōgun's household. He died on the 24th day of the 10th month in the year 1708, at the age of sixty-six, leaving no descendents of his own blood. He was buried in a Buddhist cemetery, the Jorinji, at Ushigome in Yedo (Tōkyō), where eighty years later his tomb was rebuilt, as the inscription tells us, by mathematicians of his school.

Several stories are told of Seki, some of which throw interesting side lights upon his character. One of these relates that he one time journeyed from Yedo to Kōfu, a city in Kōshū, or the Province of Kai, on a mission from his lord. Traveling in a palanquin he amused himself by noting the directions and distances, the objects along the way, the elevations and depressions, and all that characterized the topography of the region, jotting down the results upon paper as he went. From these notes he prepared a map of the region so minutely and carefully drawn that on his return to Yedo his master was greatly impressed with the powers of description of one who traveled like a *samurai* but observed like a geographer.

Another story relates how the Shōgun, who had been the Lord of Kōshū, once upon a time decided to distribute equal portions of a large piece of precious incense wood among the members of his family. But when the official who was to cut the wood attempted the division he found no way of meeting his lord's demand that the shares should be equal. He therefore appealed to his brother officials who, with one accord, advised him that no one could determine the method of cutting the precious wood save only Seki. Much relieved, the official appealed to "The Arithmetical Sage," and not in vain.

It is also told of Seki that a wonderful clock was sent from the Emperor of China as a present to the Shōgun, so arranged that the figure of a man would strike the hours. And after some years a delicate spring became deranged, so that the figure would no longer strike the bell. Then were called in the most skilful artisans of the land, but none was able to repair the clock, until Seki heard of his master's trouble. Asking that he might take the clock to his own home, he soon restored it to the Shōgun successfully repaired and again correctly striking the hours.

Such anecdotes have some value in showing the acumen and versatility of the man, and they explain why he should have been sought for a post of such responsibility as that of examiner of accounts.

The name of Seki has long been associated with the *yenri*, a form of the calculus that was possibly invented by him ⌊…⌋. It is with greater certainty that he is known for his *tenzan* method, an algebraic system that improved upon the method of the "Celestial element" inherited from the Chinese, for the *Yendan jutsu*, a scheme by which the treatments of equations and other branches of algebra is simpler than by the methods inherited from China and improved by such Japanese writers as Isomura and Sawaguchi, and for his work in determinants that antedated what has heretofore been considered the first discovery, namely the investigations of Leibniz.

As to his works, it is said that he left hundreds of unpublished man- uscripts, but if this be true most of them are lost. He also published the *Hatsubi Sampō* in 1674. In this he solved the fifteen problems given in Sawaguchi's *Kokon Sampoō-ki* of 1670, only the final equations being given.

As to Seki's real power, and as to the justice of ranking him with his great contemporaries of the West, there is much doubt. He certainly improved

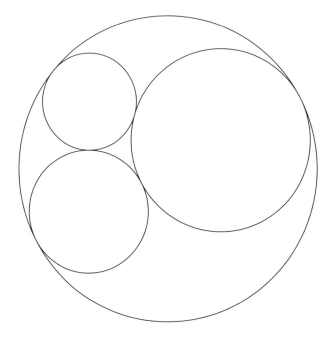

Figure 7.2. In a circle three other circles are inscribed ....

the methods used in algebra, but we are not at all sure that his name is properly connected with the *yenri*.

For this reason, and because of his fame, it has been thought best to enter more fully into his work than into that of any of his predecessors, so that the reader may have before him the material for independent judgment.

First it is proposed to set forth a few of the problems that were set by Sawaguchi, with Seki's equations and with one of Takebe's° solutions.

Sawaguchi's first problem is as follows: "In a circle three other circles are inscribed as here shown (see Figure 7.2), the remaining area being 120 square units. The common diameter of the two smallest circles is 5 units less than the diameter of the one that is next in size. Required: to compute the diameters of the various circles."

Seki solves the problem as follows: "Arrange the 'celestial element,'° taking it as the diameter of the smallest circles. Add to this the given quantity and the result is the diameter of the middle circle. Square this and call the result *A*.

"Take twice the square of the diameter of the smallest circles and add this to A, multiplying the sum by the moment of the circumference.[1] Call this product B.

"Multiply 4 times the remaining area by the moment of diameter.[2]

"This being added to B the result is the product of the square of the diameter of the largest circle multiplied by the moment of circumference. This is called C.[3]

"Take the diameter of the smallest circle and multiply it by A and by the moment of the circumference. Call the result D.[4]

"From four times the diameter of the middle circle take the diameter of the smallest circle, and from C times this product take D. The square of the remainder is the product of the square of the sum of four times the diameter of the middle circle and twice the diameter of the smallest circle, the square of the diameter of the middle circle, the square of the moment of circumference, and the square of the diameter of the largest circle. Call this X.[5]

"The sum of four times the diameter of the middle circle and twice the diameter of the smallest circle being squared, multiply it by A and by C

---

[1] By the "moment of the circumference" is meant the numerator of the fractional value of π. This is 22 in case π is taken as $\frac{22}{7}$.

[2] "Moment of diameter" means the denominator of the fractional value of π. In the case of $\frac{22}{7}$, this is 7. That is, we have 7×120.

[3] Thus far the solution is as follows: Let x = the diameter of the smallest circle, and y = the diameter of the largest circle. Then $x + 5$ is the diameter of the so-called "middle circle."

Then
$$x^2 + 10x + 25 = A,$$
$$22(3x^2 + 10x + 25) = B,$$
$$\text{and } 7 \cdot 4 \cdot 120 + B = C = 22y^2, \text{ where } \pi = \frac{22}{7}.$$

That the formula for C is correct is seen by substituting for 120 the difference in the areas as stated. We then have

$$7 \cdot 4 \cdot \frac{22}{7} \left\{ \frac{y^2}{4} - \frac{(x+5)^2}{4} - \frac{2x^2}{4} \right\} + B = C,$$

or $22(y^2 - x^2 - 10x - 25 - 2x^2 + 3x^2 + 10x + 25) = C$, or $22y^2 = C$, which is, as stated in the rule, "the product of the square of the diameter of the largest circle multiplied by the moment of circumference."

[4] I.e., $22x(x^2 + 10x + 25) = D$.

[5] I.e., $\{C[4(x + 5) - x] - d\}^2 = X$.

and by the moment of circumference.[6] This quantity being canceled with
X we get an equation of the 6th degree.[7] Finding the root of this equation
according to the reversed order we have the diameter of the smallest circle.

"Reasoning from this value the diameters of the other circles are obtained."

It may add to an appreciation or an understanding of the mathematics
of this period if we add Takebe's analysis.

Let x be the diameter of the largest circle, y that of the middle circle, and
z that of the smallest circles.

Then let $AC = a$, $AD = b$, $AB = c$, and $BC = d$, these being auxiliary
unknowns at the present time.

Then

$$2a = -z + x,$$

and

$$4a^2 = z^2 - 2zx + x^2$$

or

$$4a^2 - z^2 = -2zx + x^2.$$

Therefore

$$4b^2 = -2zx + x^2 \tag{7.1}$$

If we take y from x we have $-y + x$, which is $2c$.
Squaring,

$$4c^2 = y^2 - 2yx + x^2. \tag{7.2}$$

To y add z and we have

$$2d = y + z.$$

Squaring,

$$4d^2 = y^2 + 2yz + z^2.$$

---

[6] I.e., $22 \cdot 22y^2(x + 5)^2[2(x + 5) + 2x]^2$. This is merely the second part of the preceding
paragraph stated differently.

[7] I.e., $X = 22^2(3xy^2 + 5y^2 - x^2)^2$, and this quantity equals $22y^2(x + 5)^2(6x + 20)^2$. Their
difference is a sextic.

Subtracting z^2,[8] we have

$$4(b+c)^2 = y^2 + 2yz.$$

Subtract from this (1) and (2) and we have

$$b \times 8c = 2yz + (2z + 2y)x - 2x^2.$$

Dividing by 2,

$$b \times 4c = yz + (z+y)x - x^2.$$

Squaring,

$$b^2 \times 16c^2 = y^2z^2 + (2y^2z + 2yz^2)x$$

$$+ (y+z)^2x^2 - (2y + 2z)x^3 + x^4. \qquad (7.3)$$

Multiplying (7.1) by (7.2) we also have

$$b^2 \times 16c^2 = -2y^2zx + (y^2 + 4yz)x^2 - (2y + 2z)x^3 + x^4,$$

which being canceled with the expression in (7.3) gives

$$y^2z^2 + (4y^2z + 2yz^2)x + (-4yz + z^2)x = 0,$$

from which, by canceling z,

$$y^2z + (4y^2 + 2yz)x + (-4y + z)x^2 = 0.$$

This may be written in the form

$$y^2z + (x^2z - 4x^2y) + (4y^2 + 2yz)x = 0.$$

Takebe has now eliminated his auxiliary unknowns, and he directs that the quantity in the first parenthesis be squared and canceled with the square of the rest of the expression,[9] and that the rest of the steps be followed as in Seki's solution. In this he expresses y and z in terms of x and given quantities and thus finds an equation of the sixth degree in x.

---

[8] And noting that $d^2 - (\frac{1}{2})z^2 = (b+c)^2$.

[9] This amounts to equating $x^2z - 4x^2y$ to $-[y^2z + (4y^2 + 2yz)x]$, and then squaring and canceling out like terms.

Without attempting to carry out his suggestions, enough has been given to show his ingenuity in elimination.

Notes

Yoshida, Imamura, Isomura, Muramatsu, and Sawaguchi: Japanese mathematicians active in roughly the first half of the seventeenth century.
Takebe: Takebe Hikojirō Kenkō was active in the early part of the eighteenth century.
"celestial element": This may be understood as what we would call the unknown quantity.

"Her Absolute, Incomparable Uniqueness"

B. L. van der Waerden, 1935

Emmy Noether (1882–1935) worked in several areas of mathematics, including most notably abstract algebra. Hermann Weyl, in his memorial address, described her as "a great mathematician, the greatest, I firmly believe, that her sex has ever produced." The passages reproduced here are taken from a translation of her 1935 obituary by Bartel Leendert van der Waerden, a Dutch mathematician and historian of mathematics.

B. L. van der Waerden (1903–1996), trans. Christina M. Mynhardt, "Obituary of Emmy Noether," in J. W. Brewer and Martha K. Smith (eds.), *Emmy Noether: A Tribute to Her Life and Work* (New York, 1981), pp. 93–94, 97–98. Reprinted by permission.

Fate has tragically taken from our science a very important, entirely unique personality. Our faithful journal co-worker Emmy Noether died on April 14, 1935 as a result of an operation. She was born in Erlangen on March 23, 1882, the daughter of the well-known mathematician Max Noether.

Her absolute, incomparable uniqueness cannot be explained by her outward appearance only, however characteristic this undoubtedly was. Her individuality is also by no means exclusively a consequence of the fact that she was an extremely talented mathematician, but lies in the whole structure of her creative personality, in the style of her thoughts, and the goal of her will. For as these thoughts were primarily mathematical thoughts and the will primarily intent on scientific recognition, so must we first analyze her mathematical accomplishments if we want to understand her personality at all.

One could formulate the maxim by which Emmy Noether always let herself be guided as follows: *All relations between numbers, functions, and operations become clear, generalizable, and truly fruitful only when they are separated from their particular objects and reduced to general concepts.* For her this guiding principle was by no means a result of her experience with the importance of scientific methods, but an a priori fundamental principle of her thoughts. She could conceive and assimilate no theorem or proof before it had been abstracted and thus made clear in her mind. She could think only in concepts, not in formulas, and this is exactly where her strength lay. In this way she was forced by her own nature to discover those concepts that were suitable to serve as bases of mathematical theories.

⌀

Indefatigable and in spite of all external difficulties, she proceeded along the way indicated by the concepts she formed. Also, when she lost her teaching rights in Göttingen in 1933 and was appointed to the women's college in Bryn Mawr (Pennsylvania), she soon gathered a school around her there and in nearby Princeton. Her research, which covered commutative algebra, commutative arithmetic, and noncommutative algebra, now turned to noncommutative arithmetic, but was suddenly terminated by her death.

As characteristic features we have found: An exceptionally energetic and consistent pursuit of abstract elucidation of the material to complete methodological clarity; a stubborn clinging to methods and concepts once they had been acknowledged as being correct, even when they still appeared to her contemporaries as abstract and futile; an aspiration to classify all special relationships under specific general abstract models.

Indeed, her thoughts deviated in some respects from those of most other mathematicians. We are all so dependent on figures and formulas. For her these resources were useless, rather annoying. Her sole concern was with concepts, not with intuition or calculations. The German letters which she scribbled down hurriedly on the blackboard or on paper in typical simplified form were for her representations of concepts, not objects of a more or less mechanical calculation.

This totally unintuitive and unanalytical attitude was undoubtedly also one of the main causes of the complexity of her lectures. She had no didactical gifts, and the great pains she took to explain her remarks by quickly spoken interjections even before she had finished speaking were more likely to have the opposite effect. And still how exceptionally great was the impact of her talks, everything notwithstanding! The small, faithful audience, mostly consisting of a few advanced students and often just as many lecturers and foreign guests, had to exert themselves to the utmost to keep up. When that was done, however, one had learned far more than from the most excellent lecture. Completed theories were almost never presented, but usually those that were still in the making. Each of her lecture series was a paper. And nobody was happier than she herself when such a paper was completed by her students. Completely unegotistical and free of vanity, she never claimed anything for herself, but promoted the works of her students above all. For all of us she always wrote the introductions in which the main ideas of our work, which we initially never could understand and express in such clarity on our own, were explained. She was a faithful friend to us and at the same time a strict and unprejudiced judge. As such she was also invaluable to *Mathematische Annalen.*°

As mentioned earlier, her abstract nonintuitive concepts initially found little acknowledgement. As the success of her methods also became clear to those of a different mind, this situation changed accordingly, and during the last eight years, prominent mathematicians from home and abroad went to Göttingen to ask her advice and listen to her lectures. In 1932 she shared the Ackermann-Teubner commemorative prize for arithmetic and algebra with E. Artin.° And throughout the world today the triumphant progress of modern algebra which developed from her ideas seems to be unending.

Notes

Mathematische Annalen: founded in 1868, a journal with wide coverage in modern mathematics.

E. *Artin*: Emil Artin (1898–1962), Austrian mathematician whose major work was in the branch of abstract algebra dealing with the mathematical structures called rings.

"One of Your Calculating Fits"

George Bernard Shaw, 1939

Isaac Newton, that "good and great man" (see earlier extract), has been the subject of a number of plays, among them than this one, relatively little known, by George Bernard Shaw. It features an unlikely procession of historical figures through

Newton's drawing room and a good deal of dialogue attempting to educate us about their political problems. By contrast with more recent dramatic depictions of Newton, which have tended toward the psychologically sensational, *Good King Charles* represents him as something close to an old-fashioned dotty professor, an image possibly influenced by the image of Albert Einstein, a very prominent contemporary at the time this was written.

George Bernard Shaw (1856–1950), *In Good King Charles's Golden Days: A True History That Never Happened* (London, 1939; collected edition, 1946), pp. 163–166. Reproduced with the permission of the Society of Authors, on behalf of the estate of Bernard Shaw.

Newton, aged 38, comes in from the garden, hatless, deep in calculation, his fists clenched, tapping his knuckles together to tick off the stages of the equation. He stumbles over the mat.

MRS BASHAM. Oh, do look where youre° going, Mr Newton. Someday youll walk into the river and drown yourself. I thought you were out at the university.

NEWTON. Now dont scold, Mrs Basham, dont scold. I forgot to go out. I thought of a way of making a calculation that has been puzzling me.

MRS BASHAM. And you have been sitting out there forgetting everything else since breakfast. However, since you have one of your calculating fits on I wonder would you mind doing a little sum for me to check the washing bill. How much is three times seven?

NEWTON. Three times seven? Oh, that is quite easy.

MRS BASHAM. I suppose it is to you, sir; but it beats me. At school I got as far as addition and subtraction; but I never could do multiplication or division.

NEWTON. Why, neither could I: I was too lazy. But they are quite unnecessary: addition and subtraction are quite sufficient. You add the logarithms of the numbers; and the antilogarithm of the sum of the two is the answer. Let me see: three times seven? The logarithm of three must be decimal four seven seven or thereabouts. The logarithm of seven is, say, decmial eight four five. That makes one decimal three two two, doesnt it? What's the antilogarithm of one decimal three two two? Well, it must be less than twentytwo and more than twenty. You will be safe if you put it down as—

Sally returns.

SALLY. Please, maam, Jack says it's twentyone.

NEWTON. Extraordinary! Here was I blundering over this simple problem for a whole minute; and this uneducated fish hawker solves it in a flash! He is a better mathematician than I.

MRS BASHAM. This is our new maid from Woolsthorp, Mr Newton. You havent seen her before.

NEWTON. Havent I? I didnt notice it. [*To Sally*] Youre from Woolsthorp, are you? So am I. How old are you?

SALLY. Twentyfour, sir.

NEWTON. Twentyfour years. Eight thousand seven hundred and sixty days. Two hundred and ten thousand two hundred and forty hours. Twelve million six hundred and fourteen thousand, four hundred minutes. Seven hundred and fiftysix million eight hundred and sixtyfour thousand seconds. A long long life.

MRS BASHAM. Come now, Mr. Newton: you will turn the child's head with your figures. What can one do in a second?

NEWTON. You can do, quite deliberately and intentionally, seven distinct actions in a second. How do you count seconds? Hackertybackertyone, Hackertybackertytwo, Hackertybackertythree and so on. You pronounce seven syllables in every second. Think of it! This young woman has had time to perform more than five thousand millions of considered and intentional actions in her lifetime. How many of them can you remember, Sally?

SALLY. Oh, sir, the only one I can remember was on my sixth birthday. My father gave me sixpence: a penny for every year.

NEWTON. Six from twentyfour is eighteen. He owes you one and sixpence. Remind me to give you one and sevenpence on your next birthday if you are a good girl. Now be off.

SALLY. Oh, thank you, sir. [*She goes out*].

NEWTON. My father, who died before I was born, was a wild, extravagant, weak man: so they tell me. I inherit his wildness, his extravagance, his weakness, in the shape of a craze for figures of which I am most heartily ashamed. There are so many more important things to be worked at: the transmutations of matter, the elixir of life, the magic of light and color, above all, the secret meaning of the

Scriptures. And when I should be concentrating my mind on these I find myself wandering off into idle games of speculation about numbers in infinite series, and dividing curves into indivisibly short triangle bases. How silly! What a waste of time, priceless time!

MRS BASHAM. There is a Mr Rowley going to call on you at half past eleven.

NEWTON. Can I never be left alone? Who is Mr Rowley? What is Mr Rowley?

MRS BASHAM. Dressed like a nobleman. Very tall. Very dark. Keeps a lackey. Has a pack of dogs with him.

NEWTON. Oho! So that is who he is! They told me he wanted to see my telescope. Well, Mrs Basham, he is a person whose visit will be counted a great honor to us. But I must warn you that just as I have my terrible weakness for figures Mr Rowley has a very similar weakness for women; so you must keep Sally out of his way.

MRS BASHAM. Indeed! If he tries any of his tricks on Sally I shall see that he marries her.

NEWTON. He is married already. [*He sits at the table.*]

MRS BASHAM. Oh! That sort of man! The beast!

NEWTON. Shshsh! Not a word against him, on your life. He is privileged.

MRS BASHAM. He is a beast all the same!

NEWTON [*opening the Bible*] One of the beasts in the Book of Revelation, perhaps. But not a common beast.

MRS BASHAM. Fox the Quaker,° in his leather breeches, had the impudence to call.

NEWTON. [*interested*] George Fox? If he calls again I will see him. Those two men ought to meet.

MRS BASHAM. Those two men indeed! The honor of meeting you ought to be enough for them, I should think.

NEWTON. The honor of meeting me! Dont talk nonsense. They are great men in their very different ranks. I am nobody.

MRS BASHAM. You are the greatest man alive, sir. Mr Halley° told me so.

NEWTON. It was very wrong of Mr Halley to tell you anything of the sort. You must not mind what he says. He is always pestering me to publish my methods of calculation and to abandon my serious studies. Numbers! Numbers! Numbers! Sines, cosines, hypotenuses, fluxions, curves small enough to count as straight lines, distances between two points that are in the same place! Are these philosophy? Can they make a man great?

He is interrupted by Sally, who throws open the door and announces visitors.

SALLY. Mr Rowley and Mr Fox.

King Charles the Second, aged 50, appears at the door, but makes way for George Fox the Quaker, a big man with bright eyes and a powerful voice in reserve, aged 56. He is decently dressed but his garments are made of leather.

CHARLES. After you, Mr Fox. The spiritual powers before the temporal.
FOX. You are very civil, sir; and you speak very justly. I thank you [*he passes in*].

Sally, intensely impressed by Mr Rowley, goes out.

FOX. Am I addressing the philosopher Isaac Newton?
NEWTON. You are, sir.

Notes

youre: Shaw's practice was not to mark contractions with apostrophes.
Fox the Quaker: George Fox (1624–1691), one of the founders of the Religious Society of Friends (Quakers).
Mr Halley: Edmond Halley (1656–1742), an early supporter of Newton.

Analysis Incarnate

Carl Boyer, 1968

During the twentieth century the discipline of the history of mathematics burgeoned, and several textbooks were produced which remain well-loved classics. One is this, Carl Boyer's 1968 *A History of Mathematics*, which the author intended as a textbook in which the history of mathematics was presented "with fidelity, not only to mathematical structure and exactitude, but also to historical perspective and detail." This passage about Leonhard Euler exemplifies the attractive blend which results.

Carl B. Boyer (1906–1976), *A History of Mathematics* (New York, 1968), pp. 481–488. Copyright © 1968 by John Wiley & Sons, Inc. Reproduced by permission of John Wiley & Sons, Inc.

The history of mathematics during the modern period is unlike that of antiquity or the medieval world in at least one respect: no national group remained the leader for any prolonged period. In ancient times Greece stood head and shoulders over all other peoples in mathematical achievement; during much of the Middle Ages the level of mathematics in the Arabic world was higher than elsewhere. From the Renaissance to the eighteenth century the center of mathematical activity had shifted repeatedly—from Germany to Italy to France to Holland to England. Had religious persecution not driven the Bernoulli family from Antwerp, Belgium might have had its turn; but the family emigrated to Basel, and as a result Switzerland was the birthplace of many of the leading figures in the mathematics of the early eighteenth century. We have already mentioned the work of four of the mathematicians of the Bernoulli clan, as well as that of Hermann, one of their Swiss protégés. But the most significant mathematician to come from Switzerland during that time—or any time— was Leonhard Euler (1707–1783), who was born at Basel.

Euler's father was a clergyman who, like Jacques Bernoulli's father, hoped that his son would enter the ministry. However, the young man studied under Jean Bernoulli and associated with his sons, Nicolaus and Daniel, and through them discovered his vocation. The elder Euler also was adept in mathematics, having been a pupil under Jacques Bernoulli, and helped to instruct the son in the elements of the subject, despite his hope that Leonhard would pursue a theological career. At all events, the young man was broadly trained, for to the study of mathematics he added theology, medicine, astronomy, physics, and oriental languages. This breadth stood him in good stead when in 1727 he heard from Russia that there was an opening in medicine in the St. Petersburg Academy, where the young Bernoullis had gone as professors of mathematics. This important institution had been established only a few years earlier by Catherine I along lines laid down by her late husband, Peter the Great, with the advice of Leibniz. On the recommendation of the Bernoullis, two of the brightest luminaries in the early days of the academy, Euler was called to be a member of the section on medicine and physiology; but on the very day that he arrived in Russia, Catherine died. The fledgling Academy very nearly succumbed with her, because the new rulers showed less sympathy for learned foreigners than had Peter and Catherine.

The Academy somehow managed to survive, and Euler, in 1730, found himself in the chair of natural philosophy rather than in the medical section. His friend Nicolaus Bernoulli had died, by drowning, in St. Petersburg the year before Euler arrived, and in 1733 Daniel Bernoulli left Russia to occupy the chair in mathematics at Basel. Thereupon Euler at the age of twenty-six became the Academy's chief mathematician. He married and settled down to pursue in earnest mathematical research and raise a family that ultimately included thirteen children. The St. Petersburg Academy had established a research journal, the *Comentarii Academiae Scientiarum Imperialis Petropolitanae*, and almost from the start Euler contributed a spate of mathematical articles. The editor did not have to worry about a shortage of material as long as the pen of Euler was busy. It was said by the French academician François Arago that Euler could calculate without any apparent effort, "just as men breathe, as eagles sustain themselves in the air." As a result, Euler composed mathematical memoirs while playing with his children. In 1735 he had lost the sight of his right eye—through overwork, it is said—but this misfortune in no way diminished the rate of output of his research. He is supposed to have said that his pencil seemed to surpass him in intelligence, so easily did memoirs flow, and he published more than 500 books and papers during his lifetime. For almost half a century after his death, works by Euler continued to appear in the publications of the St. Petersburg Academy. A bibliographical list of Euler's works, including posthumous items, contains 886 entries; and it is estimated that his collected works, now being published under Swiss auspices, will run close to seventy-five substantial volumes. His mathematical research during his lifetime averaged about 800 pages a year; no mathematician has ever exceeded the output of this man whom Arago characterizes as "Analysis Incarnate."

Euler early acquired an international reputation; even before leaving Basel he had received an honorable mention from the Parisian Académie des Sciences for an essay on the masting of ships. In later years he frequently entered essays in the contests set by the Académie, and twelve times he won the coveted biennial prize. The topics ranged widely, and on one occasion, in 1724, Euler shared with Maclaurin and Daniel Bernoulli a prize for an essay on the tides. (The Paris prize was won twice by Jean Bernoulli and ten times by Daniel Bernoulli.) Euler was never guilty of false pride, and he wrote works on all levels, including textbook material for use in the Russian schools. He generally wrote in Latin, and sometimes in French, although German was his native tongue. Euler had an unusual language facility, as one should expect of a person with a Swiss background. This was fortunate, for one of the distinguishing marks of eighteenth-century mathematics was the readiness with which scholars moved from one country to another, and here Euler encountered no language problems. In 1741 Euler was invited by Frederick the Great to join the Berlin Academy, and the invitation was accepted. (Jean and Daniel Bernoulli also were invited from Switzerland, but they declined.) Euler spent twenty-five years at Frederick's court, but during this period he continued to receive a pension from Russia, and he submitted numerous papers to the St. Petersburg Academy, as well as to the Prussian Academy.

Euler's stay at Berlin was not entirely happy, for Frederick preferred a scholar who scintillated, as did Voltaire. The monarch, who valued philosophers above geometers, referred to the unsophisticated Euler as a "mathematical cyclops," and relationships at the court became intolerable for Euler. Catherine the Great was only too eager to have the prolific mathematician resume his place in the St. Petersburg Academy, and in 1788 Euler returned to Russia. During this year Euler learned that he was losing by cataract the sight of his remaining eye, and he prepared for ultimate blindness by practicing writing with chalk on a large slate and by dictating to his children. An operation was performed in 1771, and for a few days Euler saw once more; but success was short-lived and Euler spent almost all of the last seventeen years of his life in total darkness. Even this tragedy failed to stem the flood of his research and publication, which continued unabated until in 1783, at the age of seventy-six, he suddenly died while sipping tea and enjoying the company of one of his grandchildren.

From 1727 to 1783 the pen of Euler had been busy adding to knowledge in virtually every branch of pure and applied mathematics, from the most elementary to the most advanced. Moreover, in most respects Euler wrote in the language and notations we use today, for no other individual was so largely responsible for the form of college-level mathematics today as was Euler, the most successful notation-builder of all times. ⌐...⌐ The definitive use of the Greek letter π for the ratio of circumference to diameter in a circle also is largely due to Euler, although a prior occurrence is found in 1706, the year before Euler was born—in the *Synopsis Palmariorum Matheseos, or A New Introduction to the Mathematics*, by William Jones (1675–1749). It was Euler's adoption of the symbol π in 1737, and later in his many popular textbooks, that made it widely known and used. ⌐...⌐

It is not only in connection with designations for important numbers that today we use notations introduced by Euler. In geometry, algebra, trigonometry, and analysis we find ubiquitous use of Eulerian symbols, terminology, and ideas. The use of the small letters a, b, c for the sides of a triangle and of the corresponding capitals A, B, C for the opposite angles stems from Euler, as does the application of the letters r, R, and s for the radius of the inscribed and circumscribed circles and the semiperimeter of the triangle respectively. The beautiful formula $4rRs = abc$ relating the six lengths also is one of the many elementary results attributed to him, although equivalents of this result are implied by ancient geometry. The designation lx for logarithm of x, the use of the now-familiar $\sum$ to indicate a summation, and, perhaps most important of all, the notation $f(x)$ for a function of x (used in the Petersburg *Commentaries* for 1734–1735) are other Eulerian notations related to ours. Our notations today are what they are more on account of Euler than of any other mathematician in history.

⌐⌐

The first volume of the *Introductio*° is concerned from start to finish with infinite processes—infinite products and infinite continued fractions, as well as innumerable infinite series. In this respect the work is the natural generalization of the views of Newton, Leibniz, and the Bernoullis, all of whom were fond of infinite series. However, Euler was very incautious in his use of such series. Although upon occasion he warned against the risk in working with divergent series,° he himself used the binomial series $1/(1 - x) = 1+x+x^2+x^3+\ldots$ for values of $x \geq 1$. In fact, by combining the two series $x/(1 - x) = x + x^2 + x^3 + \ldots$ and $x/(x - 1) = 1 + 1/x + 1.x^2 + \ldots$ Euler concluded that $\ldots 1/x^2 + 1/x + 1 + x + x^2 + x^3 + \ldots = 0$.

Despite his hardihood, through manipulations of infinite series Euler achieved results that had baffled his predecessors. Among these was the summation of the reciprocals of the perfect squares—$1/1^2 + 1/2^2 + 1/3^2 + 1/4^2 + \ldots$. Oldenburg,° in a letter to Leibniz in 1673, had asked for the sum of this series, but Leibniz failed to answer; in 1689 Jacques Bernoulli had admitted his own inability to find the sum. Euler began with the familiar series $\sin z = z - z^3/3! + z^5/5! - z^7/7! + \ldots$. Then $\sin z = 0$ can be thought of as the infinite polynomial equation $0 = 1 - z^2/3! + z^4/5! - z^6/7! + \ldots$ (obtained by dividing through by z), or, if z^2 is replaced by w, as the equation $0 = 1 - w/3! + w^2/5! - w^3/7! + \ldots$. From the theory of algebraic equations it is known that the sum of the reciprocals of the roots is the negative of the coefficient of the linear term—in this case $1/3!$. Moreover, the roots of the equation in z are known to be π, 2π, 3π, and so on; hence the roots of the equation in w are π^2, $(2\pi)^2$, $(3\pi)^2$, and so on. Therefore

$$\frac{1}{6} = \frac{1}{\pi^2} + \frac{1}{(2\pi)^2} + \frac{1}{(3\pi)^2} + \ldots \text{ or } \frac{\pi^2}{6} = \frac{1}{1^2} + \frac{1}{2^2} + \frac{1}{3^2} + \ldots$$

Through this carefree application to polynomials of infinite degree of algebraic rules valid for the finite case Euler had achieved a result that had baffled the older Bernoulli brothers; Euler in later years repeatedly made discoveries in similar fashion. When Jean Bernoulli learned of Euler's triumph, he wrote:

And so is satisfied the burning desire of my brother who, realizing that the investigation of the sum was more difficult than anyone would have thought, openly confessed that all his zeal had been mocked. If only my brother were alive now.

Euler's summation of the reciprocals of the squares of the integers seems to date from about 1736, and it is likely that it was to Daniel Bernoulli that he promptly communicated the result. His interest in such series always was strong, and in later years he published the sums of the reciprocals of other powers of the integers. Using the cosine series instead of the sine series, Euler similarly found the result

$$\frac{\pi^2}{8} = \frac{1}{1^2} + \frac{1}{3^2} + \frac{1}{5^2} + \ldots$$

hence the corollary summation

$$\frac{\pi^2}{12} = \frac{1}{1^2} - \frac{1}{2^2} + \frac{1}{3^2} - \frac{1}{4^2} + \ldots.$$

Many of these results appeared also in the *Introductio* of 1748, including the sums of reciprocals of even powers from $n = 2$ through $n = 26$. The series of reciprocals of odd powers are so intractable that it still is not known whether or not the sum of the reciprocals of the cubes of the positive integers is a rational multiple of π^3, whereas Euler knew that for the 26th power the sum of the reciprocals is

$$\frac{2^{24}.76977927\pi^{26}}{1.2.3\ldots 27}.$$

Notes

Introductio: Euler's *Introductio in analysin infinitorum* (Lausanne, 1748), "Introduction to the analysis of infinities."

divergent series: a series whose sum becomes larger without limit as the number of terms increases.

Oldenburg: Henry Oldenburg (c. 1619–1677) was secretary to the Royal Society from its inception until his death, developing a very wide scientific correspondence.

Hardy and Littlewood Rummage

Robert Kanigel, 1991

Srinivasa Ramanujan (1887–1920), one of the great mathematicians of the twentieth century, came to prominence when, as an unknown young man, he wrote to the famous British pure mathematician G. H. Hardy (1877–1947) at Cambridge University. The papers he sent included profound new results in number theory and infinite series. Kanigel's biography of Ramanujan, as well as giving a flavor of a modern tradition of mathematical biography which does not shy away from technical details, provides a vivid picture of one of the great dramatic set pieces in the history of mathematics: Hardy and his colleague Littlewood reading over the "Indian clerk's" mathematics for the first time.

Robert Kanigel (dates unknown), *The Man Who Knew Infinity* (London, 1991), pp. 165–169. Reprinted by permission of the author.

About nine o'clock, ⌊....⌋ they met, probably in Littlewood's rooms, and soon the manuscript lay stretched out before them. Some of the formulas were familiar while others, Hardy would write, "seemed scarcely possible to believe." Twenty years later, in a talk at Harvard University, he would invite his audience into the day that had so enriched his life. "I should like you to begin," he said, "by trying to reconstruct the immediate reactions of an ordinary professional mathematician who receives a letter like this from an unknown Hindu clerk." It was a mathematical audience, so Hardy introduced them to some of Ramanujan's theorems. Like this one, on the bottom of page three:

$$\int_0^\infty \frac{1 + (\frac{x}{b+1})^2}{1 + (\frac{x}{a})^2} \cdot \frac{1 + (\frac{x}{b+2})^2}{1 + (\frac{X}{a+1})^2} \ldots dx$$

$$= \frac{1}{2\pi^{\frac{1}{2}}} \frac{\Gamma(a + \frac{1}{2})\Gamma(b + 1)\Gamma(b - a + \frac{1}{2})}{\Gamma(a)\Gamma(b + \frac{1}{2})\Gamma(b - a + 1)}$$

The elongated S-like symbol appearing on the left-hand side of this equation, and in many other equations all through the letter, was an integral sign, a notation originating with Newton's competitor Leibniz. An integral—the idea goes back to the Greeks—is essentially an addition, a sum, but one of a peculiar, precise, and, at first glance, infuriating kind.

Imagine cutting a hot dog into disclike slices. You could wind up with ten sections half an inch thick or a thousand paper-thin slices. But however thin you sliced it, you could, presumably, reassemble the pieces back into a hot dog. Integral calculus, as this branch of mathematics is called, adopts the strategy of taking an infinite number of infinitesimally thin slices and generating mathematical expressions for putting them back together again—for making them whole, or "integral." This powerful additive process can be used to determine the drag force buffeting a wing as it slices through the air, or the gravitational effects of the earth on a man-made satellite, or indeed to solve any problem where the object is to piece together the contributions of many small influences.

You don't need integral calculus to determine the area of a neat rectangular plot of farmland; you just multiply length times width. But you could use it. And you could use the same additive methods applicable to wings and satellites to calculate the area of an irregularly shaped plot where length-times-width *won't* work. Furnish the function that mathematically defines its shape, and in principle you can get its area by "integrating" it—that is, by performing the additive process in a particular, precisely defined way.

Calculus books come littered with hundreds of ways to integrate functions. And yet, pick a function at random and chances are it can't be integrated—at least not straightforwardly. With "definite integrals" like those Ramanujan offered in his letter to Hardy, however, you're offered a back-door route to a solution.

A definite integral is "definite" in that you seek to integrate the function over a definite numerical range; the little numbers at top and bottom of the elongated *S*—the ∞ and 0 in Ramanujan's equation—tell us what it is. (In other words, you mark off a piece of the farm plot whose area you want reckoned.) When you evaluate a definite integral, you don't wind up with a general algebraic formula (as you do with indefinite integrals) but, in principle, an actual number. And sometimes, by applying the right mathematical tools, you can determine this number without integrating the function first—indeed, without being *able* to integrate it at all.

Broadly, this was what Ramanujan was doing in the theorem on page 3 of his letter to Hardy and all through the section labeled "IV. Theorems on Integrals."

This particular integral, he was saying, could be represented in terms of gamma functions. (The gamma function is like the more familiar "factorial"—4!, read "four factorial," $= 4 \times 3 \times 2 \times 1$—except that it extends the idea to numbers other than integers.) Hardy figured he could prove this theorem. Later he tried, and succeeded, though it proved harder than he thought. None of Ramanujan's other integrals were trifling exercises, either, and all would wind up, years later, the object of papers devoted to them. Still, Hardy judged, these were among the *least* impressive of Ramanujan's results.

More so were the infinite series, two of which were:

$$1 - 5\left(\frac{1}{2}\right)^3 + 9\left(\frac{1 \cdot 3}{2 \cdot 4}\right)^3 - 13\left(\frac{1 \cdot 3 \cdot 5}{2 \cdot 4 \cdot 6}\right)^3 + \ldots = \frac{2}{\pi}$$

and

$$1 + 9\left(\frac{1}{4}\right)^4 + 17\left(\frac{1\cdot 5}{4\cdot 8}\right)^4 + 25\left(\frac{1\cdot 5\cdot 9}{4\cdot 8\cdot 12}\right)^4 + \ldots = \frac{2^{\frac{3}{2}}}{\pi^{\frac{1}{2}}\{\Gamma\left(\frac{3}{4}\right)\}^2}.$$

The first wasn't new to Hardy, who recognized it as going back to a mathematician named Bauer. The second seemed little different. To a layman, in fact, it and kindred ones in Ramanujan's letter might seem scarcely intimidating at all; save for pi and the gamma function they were nothing but ordinary numbers. But Hardy and others would show how these series were derived from a class of functions called hypergeometric series first explored by Leonhard Euler and Carl Friedrich Gauss and as algebraically formidable as anybody could want.

Sometime before 1910, Hardy learned later, Ramanujan had come up with a general formula, later to be known as the Dougall–Ramanujan Identity, which under the right conditions could be made to fairly spew out infinite series. Just as an ordinary beer can is made in a huge factory, the ordinary numbers in Ramanujan's series were the deceptively simple end product of complex mathematical machinery. Of course, on the day he got Ramanujan's letter, Hardy knew nothing of this. He knew only that these series formulas weren't what they seemed. Compared to the integrals, they struck him as "much more intriguing, and it soon became obvious that Ramanujan must possess much more general theorems and was keeping a great deal up his sleeve."

Some theorems in Ramanujan's letter, of course, did look comfortably familiar. For example,

If $\alpha\beta = \pi^2$, then

$$\alpha^{-\frac{1}{4}}\left(1 + 4\alpha\int_0^\infty \frac{xe^{-\alpha x^2}}{e^{2\pi x} - 1}dx\right) = \beta^{-\frac{1}{4}}\left(1 + 4\beta\int_0^\infty \frac{xe^{-\beta x^2}}{e^{2\pi x} - 1}dx\right).$$

Hardy had proved theorems like it, had even offered a similar one as a mathematical question in the *Education Times* fourteen years before. Some of Ramanujan's formulas actually went back to the days of Laplace and Jacobi a century before. Of course, it was quite something that this Indian had rediscovered them.

But now, then, what was Hardy to make of this one, which he found on the last page of Ramanujan's letter?

If

$$u = \frac{x}{1+} \frac{x^5}{1+} \frac{x^{10}}{1+} \frac{x^{15}}{1 + \ldots}, v = \frac{v^{\frac{1}{5}}}{1+} \frac{x}{1+} \frac{x^2}{1+} \frac{x^3}{1 + \ldots}$$

then

$$v^5 = u \frac{1 - 2u + 4u^2 - 3u^3 + u^4}{1 + 3u + 4u^2 + 2u^3 + u^4}.$$

This was a relationship between continued fractions, in which the compressed notation for, say, the function u actually means this:

$$u = \cfrac{x}{1 + \cfrac{x^5}{1 + \cfrac{x^{10}}{1 + \cfrac{x^{15}}{1 + \ldots}}}}$$

The publication of this result some years hence would set off a flurry of work by English mathematicians. Rogers would furnish one ten-page proof for it in 1921. Darling would explore it, too. In 1929, Watson would approach it from a different angle, trying to steer clear of the tricky mathematical terrain of theta functions. But in 1913, Hardy could make nothing of it, classing it among a group of Ramanujan's theorems which, he would write, "defeated me completely; I had never seen anything in the least like them before. A single look at them is enough to show that they could only be written down by a mathematician of the highest class." And then, in a classic Hardy flourish, he added: "They must be true because, if they were not true, no one would have the imagination to invent them."

As Hardy and Littlewood probed the theorems before them, trying to make out what they said, where they fit into the mathematical canon, and how they might be proved or disproved, they began to reach a judgment. That the Indian's mathematics was strange and individual had been evident from the start. But now they were coming to see his work as something more. It was not "individual" in the way a rebellious teenager tries to be, camouflaging his ordinariness behind bizarre dress or hair. It was much more. "There is always more in one of Ramanujan's formulae than meets the eye, as anyone who sets to work to verify those which look the easiest will soon discover," Hardy would write later. "In some the interest lies very deep, in others comparatively near the surface; but there is not one which is not curious and entertaining."

The more they looked, the more dazzled they became. "Of the theorems sent without demonstration, by this clerk of whom we had never heard," one of their Trinity colleauges, E.H. Neville, would later write, "not one could have been set in the most advanced mathematical examination in the world." Hardy would rank Ramanujan's letter as "certainly the most remarkable I have ever received," its author "a mathematician of the highest quality, a man of altogether exceptional originality and power."

And so, before midnight, Hardy and Littlewood began to appreciate that for the past three hours they had been rummaging through the papers of a mathematical genius.

❧ 8 ❧

"By Plain and Practical Rules": Mathematics at Work

A GREAT DEAL OF WRITING ABOUT MATHEMATICS IS ABOUT MATHEMATICS in use: applied mathematics, as the modern term has it. Naturally quite a lot of that material is aimed squarely at specialists—whether professionals or amateurs in the field concerned—and assumes a good deal of background knowledge about the subject to which mathematics is being applied. Much makes no sense without specific instruments or tables in hand (a tendency was for books to be tailored to the use of specific instruments and tables produced by the same author or a crony), and many deal in extreme detail with highly specific technical procedures, never intended to be interesting to the uninitiated.

Frustratingly, this means that a good many of the most frequent kinds of mathematics writing are beyond the "popular" scope of this book. The design of fortifications, for instance, was one of the most common places where mathematics was written about in the early modern period, yet I have failed to find an example which is really accessible to the general reader. On the other hand, the mathematics of hobbies, like the geometry of kite design, the computation of the strength of home brew, or the calculation of cricket averages, is seldom, if ever, mathematical enough to engage the nonenthusiast.

So this chapter samples some of the more accessible reaches of "mathematics at work," and shows how the professional uses of mathematics changed between the sixteenth and the twentieth centuries. The first four selections survey some of the breadth of what traditionally was called "practical mathematics," in military and navigational uses, and the measurement of both the large and the (relatively) small. The final four extracts are from technical works aimed at beginners from the nineteenth and twentieth centuries and discuss such things as steam engines, plumbing, and typesetting, showing a shift in the subject matter to which mathematics could be applied.

The extracts also give a strong sense of the enduring tension between specificity and generalization: between the single-use rule of thumb and the wide-ranging tools of mathematical modeling provided by algebra. We can compare this with the different styles of arithmetic teaching we saw in Chapter 2, but different writers made different decisions, and the pendulum of fashion was always in motion.

High Marshal and Camp Master

Leonard Digges, 1579

Leonard Digges was, like Robert Recorde whom we met in Chapter 1, one of the earliest writers to publish on mathematics in English. His *Tectonicon* (1556) covered measuring, calculation, and the use of instruments and was reprinted many times, right down to 1692. This passage is from his posthumous *Stratioticos*, which presented arithmetic and algebra tailored to the needs of the military man.

Leonard Digges (c. 1515–c. 1559), *An Arithmeticall Militare Treatise, named Stratioticos: Compendiously teaching the Science of Numbers, as vvell in Fractions as Integers, and so much of the Rules and Æquations Algebraicall and Arte of Numbers Cossicall, as are requisite for the Profession of a Soldiour. Together with the Moderne Militare Discipline, Offices, Lawes and Duties in euery wel gouerned Campe and Armie to be obserued: Long since attempted by Leonard Digges Gentleman, Augmented, digested, and lately finished, by Thomas Digges, his Sonne* ⌊ . . . ⌋. (London, 1579), pp. 57–58, 53–55.

Certain Questions touching the Office of the High Marshal and Camp Master

Although the form of Camps may be altered according to the diversity of Situations, in respect of Rivers, or woods that do adjoin thereto, yet for lodging both Horsemen and Footmen commodiously, readily and without confusion, there is none better than the Square. It behooveth therefore those Officers to understand first what quantity of ground sufficeth for the lodging and encamping of some certain Regiment of Horsemen and Footmen. Which known, by the Rules ensuing he shall be able to extend the same to all numbers, and to know readily upon the view of any ground what number it is able to receive both of Footmen and Horse, and accordingly to give order to inferior Officers in what sort they shall proceed to divide their ground for every Regiment.

Question

I find by experience that 1000 Horsemen will demand as much ground to encamp on as 10,000 Footmen; I find also by experience that 550 paces Square of ground will suffice to receive either of them in one main Squadron Camp. But my desire is to divide ⸤each⸥ of them into 3 Regiments, and to lodge them ⸤separately⸥ in 3 square Camps. I demand how many paces Square ⸤each⸥ of those Camps must be.

   For the number sought I set down $1x$.° That, squared, maketh $1x^2$ of ⸤square⸥ paces, the ⸤Surface⸥ content of one of the 3 Camps. Therefore shall the three Camps be $3x^2$ of ⸤square⸥ paces. But those three were contained in 302,500 ⸤square⸥ paces, for so much is the ⸤surface⸥ content of the Camp of 550 paces Square. Behold therefore your Equation:

The Equation

$$3x^2 = 302,500$$

which reduced makes

$$1x^2 = 100,833\frac{1}{3} \text{ paces.}$$

The value of ⸤x⸥ is 317 paces, and so much Square ought every one of the three Foot Camps and Horse Camps to be, for the convenient receipt of such Regiments.

Question concerning the office of the Sergeant Major

The high Marshal commandeth that the Army shall be divided into 3 like square Battalions, every Battalion to be Armed in the front with 7 Ranks of Pikes, and that these battalions making up one front upon the Enemy be ⸤flanked⸥ on either side with a sleeve of Pikes of 5 in a rank. This order being prescribed, he delivereth the Sergeant Major 18,000 Soldiers' Pikes and short weapons, commanding that the Squadrons be made as great as possibly may be of those men. It is demanded how many in every rank of the Battalions, and in what sort the Sergeant Major shall shift his weapons;

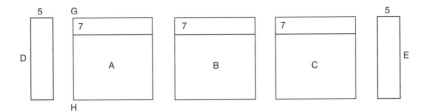

Figure 8.1. Three Battalions and two Sleeves, with seven ranks of Pikes.

how many Pikes for the sleeves, and how many of short weapons and Pikes in every Battalion ⌊separately⌋.

To resolve this Question, first I put the Marshal's order in the figure (8.1), A, B, C representing the 3 Battalions armed in the front with 7 ranks of Pikes, D and E the Sleeves, 5 in a Rank.

Now because I know not GH, the ⌊depth⌋ of the Battalion or length of the Sleeves, I suppose that ⌊to be⌋ $1x$, the which increased by 5 maketh $5x$ for one Sleeve. That, doubled, maketh $10x$ for the 2 Sleeves. Then square x; so have I x^2, the quantity of one Battalion, and so consequently $3x^2$ for the 3 Battalions. Which, adjoined to $10x$, the two Sleeves, maketh $3x^2 + 10x$ for Battalions and Sleeves.

But that should be 18,000, for so many men were delivered to make the Battalions. Behold therefore the Equation: $3x^2 + 10x = 18,000$, which reduced maketh $1x^2 + \frac{10}{3}x = 6000$, and so consequently $1x^2 = 6000 - \frac{10}{3}x$. The Value of this Root ⌊...⌋ is found to be 75 and certain fractions, which always in these Military Questions may be omitted.

I conclude, therefore, every Battalion must have 75 in a Rank, and that, multiplied by 5, maketh 375, the number of Pikes in either Sleeve. And because the Battalions are armed in the front with 7 Ranks of Pikes, I multiply 7 in 75: so have I 525, the number of Pikes in every Squadron. And that ⌊being⌋ Deducted from 5625, the Square of 75, ⌊there remains⌋ 5100, the number of short weapons in every Squadron. Thus standeth every demand resolved as followeth:

Pikes in either Sleeve: 257
Pikes in every Squadron: 525
Short weapons in every Squadron: 5625
The number of Ranks in every Squadron: 75.

Note

1x: Digges actually used "cossic" notation, in which the unknown, its square, its square root, and so on, were each denoted by a single symbol. Here modern notation has been substituted.

The Practical Gauger

William Hunt, 1673

One use of mathematics that was quite prominent in the seventeenth century was "gauging": working out the volumes of vessels containing liquids. The purpose was usually to tax the contents. Here we present a remarkably purple commendatory poem to William Hunt, a master of the art, together with part of his treatise, showing the reader how to compute the volume of liquid in various barrels. It is characteristic of this genre of writing to include some rather fantastic solid shapes as exercises, and I have excised from what follows the discussion of the baffling "cylindroid": "When the Bases are both Elliptical, but unequal, and Disproportional, or Inverted, or if one Base be a Circle, and the other an Ellipsis
. . . ."

William Hunt (fl. 1673–1739), *A guide for the practical gauger with a compendium of decimal arithmetick. Shewing briefly I. Many plain and easie ways how to gauge brewers tuns, coppers, backs, etc. also the mash-tun, either in whole, or gradually from inch to inch, with divers new tables for facilitating the work. II. The gauging of any wine, brandy, ale or oyl-cask, either in whole, or in part, with the construction and use of two tables of area's of circles, and Sybant Hantz his table of area's of segments of a circle. III. The mensuration of all manner of superficies, as board, glass, pavement, wainscot, tiling, floors, roofs, etc. also brick-work, timber and stone. Added as an appendix to the former work. Collected and published principally for the service of the farmers of his Majesties revenue of excise. By William Hunt, student in the mathematicks.* (London, 1673). Commendatory poem by Henry Coley, "Philom₍ath₎,". A8ʳ–ᵛ.

William Hunt, *The Gaugers Magazine wherein the Foundation of his Art Is briefly Explain'd and Illustrated with such Figures, As may render the Whole intelligible to a mean Capacity. By William Hunt, Gauger.* (London, 1687), pp. 223–227, 230–232.

Upon The Author And his Work

Let Criticks Carp against thy Work and thee,
Whilst I commend it, since thou art so free
Thus to discover what thou know'st in ART,
Both in the Theory and the Practick part

Of Gauging, which is now become so pure
A *Ne plus ultra*° may be put I'm sure.
How much of late years has been writ hereon?
And yet room left for thee to Build upon;
Go on! Brave Soul, still search, thy Active Pate
Is always Busied; thou regard'st not Fate,
But Aim'st at Common, more than private good,
Which shows thy veins are filled with Generous blood.
To such as in strange paths are led aside,
Thou here hast sent a Welcome pleasant Guide
For to Direct them; let them view each page
With understanding, and Attain to Gauge.
All Vessels, though of various forms they be,
Round, Square or Oval, 'tis all one to thee,
Tuns, Backs and Coolers, Coppers, and the rest,
Cum multis aliis,° work as you like best.
Here's all variety you can desire,
Return our Author thanks, and then Aspire
To understand each Problem in his Guide,
Which he with Diligence hath often tried.
Words will not Reach his Worth, Hyperbolize
I can't, nor sound his Fame up to the skies;
Yet this I'll say, and still add to his praise,
That he alone in this deserves the Bays.°

⌞Theorems⌟

Problem 118. Definition

A *Prism* is a Solid Figure contained under several Planes, two of which, being opposite, are called the Bases, and are equal, parallel, and alike situated. But the rest of the Planes are Parallelograms, in which a ⌞straight⌟ line may be everywhere applied from one Base to another, (which ⌞Base⌟ may be a Triangle, Quadrangle, Pentagon, or any other plane Surface).

Theorem

Multiply the Area of the Base ⌞...⌟ by the Altitude ⌞...⌟; the Product will be the Content.

Problem 121. Definition

A Pyramid is a Solid Figure contained under diverse Planes, set upon one ⌐...⌐ which is called the Base, from whence it decreaseth equally, less and less, till it end in a Point at the Top or Vertex. Also, in ⌐any⌐ of these Planes, a ⌐straight⌐ line may be everywhere applied from the Base to the Vertex.

Theorem

Multiply the Area of the Base (Whether it be Triangular, Quadrangular, Pentagonal, etc.) by one third part of the Altitude. The Product will be the Solid Content.

⌐Definition⌐

⌐If we cut a pyramid into two parts using a cut parallel to the base, two solid figures will remain. One will be another pyramid; the other will be what is called a frustum of a pyramid.⌐

Problem 122

Given AH and CI, the Sides at the Base⌐s⌐ of the Frustum of a ⌐square-based⌐ Pyramid, and GP the Altitude, to find the Solid Content. (See Figure 8.2.)

Theorem

As AK (the Semi-difference of AH and CI) is to CK ($= CP$, the Frustum's Altitude), so is AP ($=$ half AH the Side of the greater Base) to ZP (the Altitude of the Pyramid AZB).°

From ZP subduct GP; the Remainder is ZG, the Altitude of the Pyramid CZD.

Lastly, from AZB, the whole Pyramid, subduct the Lesser Pyramid CZD; the Remainder will be the Content of the Frustum $ABCD$.

Now to reduce the former Theorems to Practice

You must take notice that Brewers' Tuns, though of various Forms, are generally comprehended under these three, *viz.* Square, Round, and

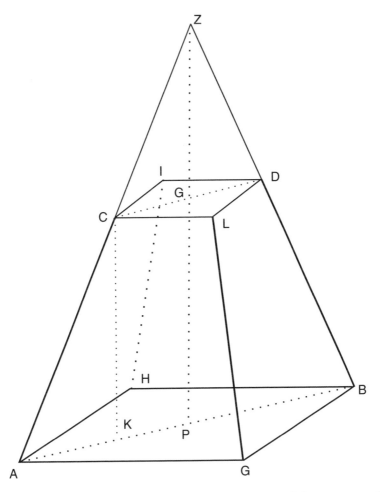

Figure 8.2. Given *AH* and *CI*, the Sides at the Bases of the Frustum of a square-based Pyramid, and *GP* the Altitude, to find the Solid Content.

Elliptical, their Sides being supposed to be straight from Top to Bottom, and their Bases parallel.

I. Of Square Tuns

1. When the Bases are both Square, and Equal, the Tun is a Prism and Gauged by Theorem 118.
2. When the Bases are both Rectangular Parallelograms, equal and alike situated, the Tun is a Prism.

3. When the Bases are both Square, but unequal, the Tun is the Frustum of a Pyramid, and Gauged by Theorem 122.

4. When the Bases are both Rectangular Parallelograms, proportional and alike situated, but unequal, the Tun is the Frustum of a Pyramid.

II. Of Round Tuns

1. When the Bases are both Circular, and equal, the Tun is a Cylinder and Gauged by Theorem 118.

2. When the Bases are both Circular, but unequal, the Tun is the Frustum of a Cone, and Gauged by Theorem 122.

III. Of Elliptical Tuns

1. When the Bases are both Elliptical, equal, and alike situated, the Tun is a Prism, and Gauged by Theorem 118.

2. When the Bases are both Elliptical, proportional, and alike situated, but unequal, the Tun is the Frustum of an Elliptical Cone, and Gauged by Theorem 122.

Notes

Ne plus ultra: literally, "no further"; so, a boundary, an outer limit.

Cum multis aliis: with many others.

the Bays: the victor's crown of laurels.

As AK …: There is some mistake here. It is evident from the diagram that K is supposed to be the foot of a line through C perpendicular to the base of the frustum. The length AK is $\frac{1}{2}(AB - CD)$, or $\frac{1}{2}(\sqrt{2}AH - \sqrt{2}CI)$, which is $\frac{1}{\sqrt{2}}(AH - CL)$. Hunt's argument is then that APZ and AKC are similar triangles, so that the length ZP can be deduced. From there it is straightforward to calculate the volumes of the pyramids CZD and AZB, and, subtracting one from the other, to find the volume of the frustum.

Geodæsia

John Love, 1688

Surveying was one of the oldest uses of mathematics ("geometry," after all, means "measuring the earth"), and it flourished in the early modern period as a practical

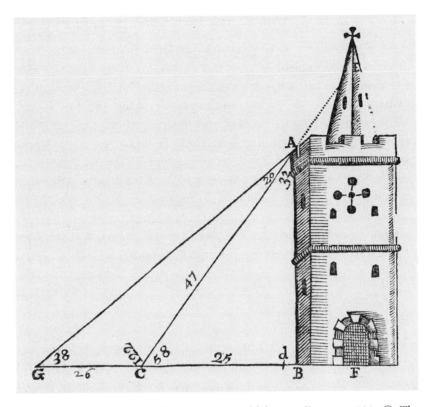

Figure 8.3. A tower whose height you would know. (Love, p. 181. © The Bodleian Libraries, University of Oxford. F 2.36 Linc.)

art which anyone could learn. New concerns with the production of accurate maps for navigation and in the colonies gave the subject a new urgency.

John Love described himself as a "philomath" ("lover of," or perhaps "friend to" mathematics), and he presumably had some professional experience as a surveyor; he does not seem to have written anything else, but this book was hugely successful, remaining in print for over a century and later appearing in special editions for the American surveyors he had envisaged on his original title page.

John Love, *Geodæsia: or, the Art of Surveying and Measuring of Land, Made Easie. Shewing, By Plain and Practical Rules, How to Survey, Protract, Cast up, Reduce or Divide any Piece of Land whatsoever; with New Tables for the ease of the Surveyor in Reducing the Measures of Land. Moreover, A more Facile and Sure Way of Surveying by the Chain, than has hitherto been Taught. As Also, How to Lay-out New Lands in America, or elsewhere: And how to make a Perfect Map of a River's Mouth or Harbour; with several other Things never yet Publish'd in our Language.* (London, 1688), pp. 180–183.

How to take the Height of a Tower, Steeple, Tree, or any such thing

Let AB be a Tower whose Height you would know (see Figure 8.3).

First, at any convenient distance, as at C, place your Semi-circle, or what other Instrument you judge more fit for the taking an Angle of Altitude, as a large Quadrant, or the like, and there observe the Angle ACB. ⌐…⌐ Measure next the distance between your Instrument and the foot of the Tower, *viz.* the line Cd, which let be 25 Yards. Then have you all the Angles given (admitting the Angle of the Tower makes with the Ground, *viz.*, d, to be the Right Angle), and the Base, to find the Perpendicular AB; which you may do, as you were taught in ⌐…⌐ Trigonometry. For if you take 58 from 90, there remains 32 for the Angle at A. Then say,

⌐As the Sine of the Angle A, 32, is to the Base Cd, 25, so is the Sine of the Angle C, 58, to the Height of the Tower, AB, or rather Ad.

So $0.530 : 25 = 0.848 : Ad$
and $Ad = 0.848 \times 25/0.530$
$= 40$ Yards.⌐°

To this 40 Yards you must add the height of your Instrument from the Ground ⌐…⌐. In this way of taking Heights, the Ground ought to be very level, or you may make great Mistakes. Also the Tower or Tree should stand perpendicular, or else you must measure to such a place where a Perpendicular would fall if let down. As, AB is not a Perpendicular, but Ad; therefore measure the Distance Cd for your Base.

This you may plainly understand by the foregoing Figure, for ⌐suppose⌐, standing at C, you were to take the Height of the Tower and Steeple to E. The Angle ECB is the same as the Angle at ACB, and if you measure only CB or Cd, you will make the Height FE the same as dA, which by the Figure you plainly perceive to be a great Error. Therefore, to take the Height FE, you should measure from C to F.

How to take the Height of a Tower, etc., when you cannot come nigh the Foot thereof

In the foregoing Figure, let AB be the Tower, and suppose CB to be a Moat, or some other hindrance, that you cannot come higher than C to take the Height. Therefore at C plant your Instrument, and take (as before) the Angle ACB, 58 degrees. Then go backwards any convenient distance, as to G, ˌandˌ there also take the Angle AGB, 38 degrees. This done, subtract 58 from 180, so have you 122 degrees, the Angle ACG. Then, 122 and 38 being taken from 180, ˌthereˌ remains 20 for the Angle GAC. The Distance GC, measured, is 26. Now by Trigonometry, say,

ˌAs the Sine of the Angle A, 20, is to the Distance GC, 26, so is the Sine of the Angle G, 38, to the Line AC.

So $0.342 : 26 = 0.616 : AC$,
and $AC = 0.616 \times 26/0.342$
$= 46.8$.

Again,
As the Sine of the right angle, B, is to the Line AC, 46.8, so is the Sine of the Angle C, 58, to the Height of the Tower Ad.

So $1 : 46.8 = 0.848 : Ad$,
and $Ad = 0.848 \times 46.8$
$= 39.7$ Yards.ˌ°

Notes

As the Sine …: Love's procedure is slightly more complex, using the logarithms of the quantities rather than the quantities themselves.

39.7 Yards: Love rounds at two different stages in the working, to produce, a little dubiously, the same result—40 yards—as in the previous calculation.

Plain Sailing

Archibald Patoun, 1762

Navigation was one of the most common uses—in print—for mathematics, certainly in the seventeenth and eighteenth centuries. We find works on the "longitude problem," of course, and books describing the use of new and more advanced instruments and tables; but there was also a basic literature on the application of spherical (or plane) geometry to course finding. So routine, indeed, was this material, that many authors were content to "borrow" one another's examples and even text, perpetuating what were often not particularly lucid ways of explaining matters.

Patoun, more original than most, was a Fellow of the Royal Society, although he does not seem to have published anything in its *Transactions*; and although it was something of a success and went to at least eight editions, this seems to have been his only book.

Archibald Patoun (dates unknown), *A Complete Treatise of Practical Navigation Demonstrated from it's First Principles: With all the Necessary Tables. To which are added, the useful Theorems of Mensuration, Surveying, and Gauging; with their Application to Practice.* (London: 6th edition, 1762), pp. 160–163.

Of Plain Sailing

This Method of Sailing supposes the Earth to be a Plane, and the Meridians parallel to one another, and likewise the Parallels of Latitude at equal Distance from one another, as they really are upon the Globe. Though this Method be in itself evidently false, yet in a short Run, and especially near the Equator, an Account of the Ship's Way may be kept by it tolerably well.

The Angle formed by the Meridian and Rhumb,° that a Ship sails upon, is called the Ship's Course. Thus, if a Ship sails on the NNE Rhumb, then her Course will be 22°, 30′. And so of others.

The distance between two Places lying on the same Parallel (counted in Miles of the Equator), or the Distance of one Place from the Meridian of another (counted as above) on the Parallel passing over that Place, is called Meridional Distance, which in Plain Sailing goes under the Name of "Departure."

Let *A* denote a certain Point on the Earth's Surface, *AC* its Meridian, and *AD* the Parallel of Latitude passing through it (see Figure 8.4). And suppose a Ship to sail from *A* on the NNE Rhumb 'til she arrive at *B*. And through *B*, draw the Meridian, *BD* (which, according to the Principles of Plain Sailing, must be parallel to *CA*), and the Parallel of Latitude, *BC*. Then the Length of *AB*, viz., how far the Ship has sailed upon the NNE Rhumb, is called her Distance. *AC* or *BD* will be her Difference of Latitude, or Northing, *CB* will be her Departure, or Easting, and the Angle *CAB* will be the Course.

Hence it is plain that the Distance sailed will always be greater than either the Difference of Latitude or ⌐the⌐ Departure, it being the Hypotenuse of a right-angled Triangle, whereof the other two are the Legs, ⌐unless⌐ the Ship sails either on a Meridian or a Parallel of Latitude. For if the Ship sails on a Meridian, then it is plain that her Distance will be just equal to her

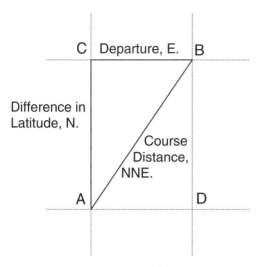

Figure 8.4. Suppose a Ship to sail from *A* on the NNE Rhumb 'til she arrive at *B*.

Difference of Latitude, and she will have no Departure; but if she sail on a Parallel, then her Distance will be the same with her Departure, and she will have no Difference of Latitude. It is evident also, from the Scheme, that if the Course be less than 4 Points, or 45 Degrees, its Complement, *viz.*, the other Oblique Angle, will be greater than 45 Degrees, and so the Difference of Latitude will be less than the Departure. And, lastly, if the Course be just 4 Points, the Difference of Latitude will be equal to the Departure.

Since the Distance, Difference of Latitude, and Departure form a right-angled Triangle, in which the Oblique Angle opposite to the Departure is the Course, and the other its Complement, therefore having any two of these given, we can find the rest. And hence arise the Cases of Plain Sailing, which are as follow.

Case 1

Course and Distance given, to find Difference of Latitude and Departure.

Example

Suppose a Ship sails from the Latitude of 30°, 25′ North, NNE, 32 Miles. Required: the Difference of Latitude and Departure, and the Latitude come to.

⚓

As the Sine of 90° is to the Distance AC, 32, so is the Sine of the Course, A: 22°, 30′, to the Departure BC.

So $1 : 32 = 0.383 : BC$,
and $BC = 0.383 \times 32$
$= 12.25.$

So the Ship has made 12.25 Miles of Departure Easterly, or has got so far to the Eastward of her Meridian. Then, for the Difference of Latitude, or Northing, the Ship has made, we have:

As the Sine of 90° is to the Distance AC, 32, so is the Cosine of the Course, A: 22°, 30′, to the Difference of Latitude, AB.

So $1 : 32 = 0.924 : AB$,
and $AB = 0.924 \times 32$
$= 29.56.°$

So the Ship has differed her Latitude, or made of Northing, 29.57 Minutes.

And since her former Latitude was North, and her Difference of Latitude also North, therefore to the Latitude sailed from, 30°, 25′N, add the Difference of Latitude, 29.57′, and the Sum is the Latitude come to: 30°, 54.57′N.

Notes

Rhumb: a path on the earth's surface at a constant compass direction.
29.56: Patoun makes it 29.57, due presumably to a rounding error.

High-Pressure Engines

William Templeton, 1833

This *Pocket Companion* has much in common with the practical manuals of a much earlier era, and William Templeton, an engineer, remained in print on subjects like "The locomotive engine popularly explained" until the early twentieth century. He was the author of mathematical tables "for practical men," but like many of his predecessors he also aimed his work at "scientific gentlemen."

At the same time, this section on the mathematics of steam engines bears witness to the fact that the nineteenth century saw any number of new technological uses for mathematics, in a development which would ultimately drive the nonspecialist away from much of this type of practical mathematics.

William Templeton (fl. 1833–1852), *The Millwright & Engineer's Pocket Companion; comprising decimal arithmetic, tables of square and cube roots, practical geometry, mensuration, strength of materials, mechanic powers, water wheels, pumps and pumping engines, steam engines, tables of specific gravities, etc.* (London: 2nd edition, 1833), pp. 142–144.

The effective power obtained by means of a high-pressure engine is nearly two-thirds of the force of the steam, one-third being expended in friction, etc. Hence, multiply the cylinder's area in inches by the force of the steam in pounds, and by the velocity of the piston in feet per minute; deduct $\frac{1}{3}$ of the product, and divide the remainder by 33,000. The quotient will be the force of the engine, expressed in horsepower.

Example

Required: the power of an engine, the cylinder being 8 inches diameter, and stroke 2 feet, the engine making 50 revolutions per minute, and the weight upon the safety valve equal ͺtoͺ 30lbs. per square inch.

$$8 \text{ inches diameter} = 50.2656 \text{ inches area, and}$$

$$50 \text{ revolutions} \times 4 \text{ feet} = 200 \text{ feet velocity.}$$

$$\text{Then, } \frac{50.2656 \times 30\text{lbs} \times 200 \times 2}{3} = \frac{201062.4}{33,000}$$

$$= 6.09 \text{ horsepower, nearly.}$$

Or ͺconverselyͺ, multiply 49,500 by the number of horsepower required; divide the product by the force of the steam in pounds, multiplied by the velocity of the piston in feet per minute, and the quotient will be the area of the cylinder.

Example

Required: the diameter of a cylinder for an engine of 12 horsepower, with, working pressure 35lbs. per square inch, length of stroke 2 feet 6 inches, and making 45 revolutions per minute.

$$\frac{49{,}500 \times 12}{45 \times 5 \times 35} = \frac{594{,}000}{7875}$$

$$= 75.43 \text{ inches area, or } 9.8 \text{ inches diameter nearly.}$$

To find an Equivalent Force of the Steam, when the Engine is working expansively

Rule 1

Divide the length of the stroke in inches by the distance (also in inches) that the piston moves before the steam is shut off. This gives the number of times the steam is expanded. Divide the pressure on the boiler in pounds by this quotient. This gives the number of pounds to which the steam is expanded.

Rule 2

Add 1 to the natural logarithm of the number of times by which the steam is expanded, and multiply the logarithm by the number of pounds to which the steam is expanded, and the product is the uniform force of the steam acting throughout the whole stroke.

Example

Let the steam in the boiler of a high-pressure engine equal 45lbs. per inch, the length of stroke 4 feet, and the steam to be shut off after the piston has moved 16 inches. Required, an equivalent force of the steam in the cylinder.

4 feet = 48 inches, and 48 ÷ 16 = 3.
Then 45 ÷ 3 = 15lbs.

Now, the natural logarithm of 3 is 1.0986123, and 1 + 1.0986123 = 2.0986123. And 2.0986123 × 15 = 31.4791845lbs., the uniform force of the steam.

The Strength of Materials

Lucius D. Gould, 1853

Sections on the strength of materials and the rules for scaling constructions up or down had long appeared in mathematical manuals. Lucius D. Gould, architect, had some new materials to work with in the nineteenth century. Gould himself is an obscure figure; this volume, in various versions, seems to have been his only book.

Like Templeton's account of high-pressure engines in the previous extract, this passage illustrates the resurgence in the nineteenth century of rules of thumb rather than general principles in works of this type, reflecting the increasing complexity of the underlying mathematical models compared with their eighteenth-century counterparts.

Lucius D. Gould (dates unknown), *The American House Carpenters- and Joiners- Assistant: Being a New and Easy System of Lines, Founded on Geometrical Principles. For Cutting every Description of Joints. And for Framing the most Difficult Roof. To which is added A Complete Treatise on Mathematical Instruments. Also, Mensuration, Tables of the Weights and Cohesive Strength of the Several Materials used in the Construction of Buildings, etc.* (New York, 1853), pp. 125–127.

1. To find the Breadth of a uniform Cast-Iron Beam, to bear a given weight in the middle

Rule

Multiply the length between the supports in feet, by the weight to be supported in pounds, and divide the product by 850 times the square of the depth in inches; the quotient will be the required breadth in inches.

2. To find the Depth of a uniform Cast-Iron Beam, to bear a given weight in the middle

Rule 1

Multiply the length of bearing in feet, by the weight to be supported in pounds, and divide this product by 850 times the breadth in inches; and the square root of the quotient will be the required depth in inches.

When no particular breadth or depth is determined by the nature of the situation for which the beam is intended, it will be found sometimes convenient to assign some proportion; as, for example, let the breadth be the nth part of the depth, n representing any number at will. Then the rule will be as follows:

Rule 2

Multiply n times the length in feet, by the weight in pounds; divide this product by 850, and the cube root of the quotient will be the depth required; the breadth will be the nth part of the depth.

Note

It may be remarked here that the rules are the same for inclined as for horizontal beams, when the horizontal distance between the supports is taken for the length of bearing.

Example

In a situation where the flexure of a beam is not a material defect, it is required to support a load which cannot exceed 33,600 pounds, or 15 tons, in the middle of a cast iron beam, the distance of the supports being 20 feet, and making the breadth a fourth part of the depth. In this case,

$$n = 4 \text{ and } \frac{4 \times 20 \times 33,600}{850} = 3162.35.$$

The cube root of 3162.35 is nearly 14.68 inches, the depth required; the breadth is

$$14.68 \div 4 = 3.87 \text{ inches.}$$

In practice, therefore, whole numbers should be used, and the beam be made 15 inches in depth, and 4 inches in breadth.

3. To find the Breadth of a uniform Cast-Iron Beam, when the load is not in the middle between the supports

Rule

Multiply the distance between the load and the nearest support in feet, by the distance between the load and the furthest support in feet; and four times this product, divided by the whole length between the supports, will give the effective leverage of the load in feet; this quotient being used instead of the length, in any of the foregoing Rules, will give the breadth and depth.

Example

Instead of placing 15 tons in the middle of a beam, as in the last Example, let it be placed at 5 feet from one end, and 15 feet from the other.

$$\frac{5 \times 15 \times 4}{20} = 15,$$

which is the number to be employed instead of the whole length, as in the last example; that is

$$\frac{4 \times 15 \times 33,600}{850} = 2372 \text{ nearly.}$$

And the cube root of 2372 is nearly 13.34 inches, the depth of the required beam; and

$$13.34 \div 4 = 3.33 \text{ inches}$$

gives the breadth, or the beam is nearly $13\frac{1}{2}$ inches by $3\frac{1}{2}$ inches. In the foregoing Example it was 15 inches by 4 inches.

4. To substitute Beams of Wrought Iron, of Oak, or of Yellow Fir, instead of having Beams of Cast Iron, in either of the foregoing problems

Rule

Instead of using 850 as a divisor, when the beam is of wrought iron, employ 952; when it is of oak, 212; and when it is of yellow fir, 255.

5. To find the Breadth of a uniform Cast-Iron Beam, when the load is distributed over the length of a Beam, supported at both ends

Rule

When the load is uniformly distributed over the length of a beam, it supports double the weight that it would do if the load were laid on the middle; therefore the divisor, in the preceding examples, is changed from 850 to 1700.

Example

In a situation where an arch cannot be used for want of abutments,° it is necessary to leave an opening 15 feet wide, in an 18-inch brick wall. Required: the depth of two cast-iron beams to support the walls over the openings, each beam to be 2 inches thick, and the height of the wall intended to rest on the beam being 30 feet.

The wall contains

$$30 \times 15 \times 1\frac{1}{2} = 675 \text{ cubic feet;}$$

and as a cubic foot of brickwork weighs about 100 lbs., the weight of the wall will be about 67,500 lbs.; and half this weight, or 33,750 lbs., will be the load upon one of the beams. Since the breadth is supposed to be given, the depth will be found by problem 2, if 1700 be used as the constant divisor; thus,

$$\frac{15 \times 33,750}{1700 \times 2} = 149 \text{ nearly.}$$

The square root of 149 is $12\frac{1}{4}$ nearly; therefore each beam should be $12\frac{1}{4}$ inches deep, and 2 inches in thickness. This operation gives the actual strength necessary to support the wall, but double the weight is usually taken in practice to allow for accidents. In this manner the strength for brestsummers,° lintels, and the like, may be determined.

Notes

abutments: supports which resist outward thrust.

brestsummer: "a 'summer' or beam extending horizontally over a large opening, and sustaining the whole superstructure of wall, etc.; e.g. the beam over a shop-front, the lower beam of the front of a gallery, and the like." (*OED.*)

Plumbing and Hydraulics

William H. Dooley, 1920

Authors named William H. Dooley published works on industrial education, clothing style, and shoemaking in the early twentieth century, but it is not quite clear whether any of them was the same person as the author of this volume of "vocational mathematics." Firmly in the tradition of practical mathematics addressed to practical readers stretching right back to, say, the works on "gauging" of the seventeenth century, it is concerned with mathematical procedures for solving specific practical problems in various areas. Dooley follows the nineteenth-century tendency to give practical recipes with little in the way of proof. Algebra is used only sparingly, even when it would clarify the relationships between some of the problems and reduce the amount of material needing to be rote-learned.

For a different take on air pressure and barometers, see the extracts from *The Ladies' Diary* in Chapter 3.

William H. Dooley (dates unknown), revised by A. Ritchie-Scott, *Vocational Mathematics* (London, 1920), pp. 169–172, 174–177.

Atmospheric pressure

Atmospheric pressure is often expressed as a certain number of "atmospheres." The pressure of one "atmosphere" is the weight of a column of air, one square inch in area.

At sea level the average pressure of the atmosphere is approximately 15 pounds per square inch.

The pressure of the air is measured by an instrument called a barometer. The barometer consists of a glass tube, about $31\frac{1}{2}$ inches long, which has been entirely filled with mercury (thus removing all air from the tube) and inverted in a vessel of mercury.

The space at the top of the column of mercury varies as the air pressure on the surface of the mercury in the vessel increases or decreases. The pressure is read from a graduated scale which indicates the distance from the surface of the mercury in the vessel to the top of the mercury column in the tube.

Questions

1. Four atmospheres would mean how many pounds?
2. Give in pounds the following pressures: 1 atmosphere; $\frac{1}{2}$ atmosphere; $\frac{3}{4}$ atmosphere.
3. If the air, on the average, will support a column of mercury 30 inches high with a base of 1 square inch, what is the pressure of the air? (One cubic foot of mercury weighs 849 pounds.)

Water pressure

When water is stored in a tank, it exerts pressure against the sides, whether the sides are vertical, oblique, or horizontal. The force is exerted perpendicularly to the surface on which it acts. In other words, every pound of water in a tank, at a height above the point where the water is to be used, possesses a certain amount of energy due to its position.

It is often necessary to estimate the energy in the tank at the top of a house or in the reservoir of a town or city, so as to secure the needed water pressure for use in case of fire. In such problems one must know the perpendicular height from the water level in the reservoir to the point of discharge. This perpendicular height is called the *head*.

Pressure per Square Inch

To find the pressure per square inch exerted by a column of water, multiply the head of water in feet by 0.434. The result will be the pressure in pounds.

Head

To find the *head* of water in feet, if the pressure (weight) per square inch is known, multiply the pressure by 2.31.

Lateral Pressure

To find the total lateral (sideways) pressure of water upon the sides of a tank, multiply the area of the submerged side, in inches, by the pressure due to one half the depth.

Thickness of Pipe

To find the thickness of a lead pipe necessary for a given head of water, multiply the head in feet by the size of the pipe required, expressed as a decimal, and divide this result by 750. The quotient will represent the thickness required in hundredths of an inch.

Velocity of Water

Velocity through Pipes

To calculate the velocity of water flowing through a horizontal straight pipe of given length and diameter, the head of water above the centre of the pipe being known, multiply the head of water in feet by 2500 and divide the result by the length of the pipe in feet multiplied by 13.9, divided by the inner diameter of the pipe in inches. The square root of the quotient gives the velocity in feet per second.

Let

$$l = \text{length of the pipe in feet}$$
$$d = \text{diameter of the pipe}$$
$$H = \text{head of water above the centre of pipe}$$
$$V = \text{velocity of water in feet per second}$$

Then

$$V = \sqrt{\frac{H \times 2500}{l \times \frac{13.9}{d}}}$$

Head and velocity

To find the head which will produce a given velocity of water through a pipe of a given diameter and length, multiply the square of the velocity, expressed in feet per second, by the length of pipe, multiplied by the quotient obtained by dividing 13.9 by the diameter of the pipe in inches. Divide this result by 2500, and the final quotient will give the head in feet.

$$h = \frac{V^2 \times L \times \frac{13.9}{d}}{2500}$$

V = velocity of water expressed in ft. per sec.
L = length of pipe expressed in feet
d = diameter of pipe expressed in inches

Power

Power is the time rate of doing work. The unit of power is the horsepower (H.P.), which represents 33,000 foot pounds a minute or 550 foot pounds a second. A foot pound is the amount of work necessary to lift a pound through a distance of one foot

Raising water

To find the power necessary to raise water to any given height, multiply the product of cubic feet required per minute and the number of feet through which it is to be lifted by 62.3 and divide this product by 33,000. This will give the nominal horsepower required. If the amount of water required per minute is in gallons, the multiplier should be 8.3 instead of 62.3.

Water Power

When water flows from one level to another, it exerts a certain amount of energy, which is the capacity for doing work. This energy may be utilized by such means as the water wheel, the turbine, and the hydraulic ram.

Friction, which must be considered when one speaks of water power, is the resistance which a substance encounters when moving through or over another substance. The amount of friction depends upon the pressure between the surfaces in contact.

When work is done, a part of the energy which is expended is apparently lost. In the case of water this is due to the friction. All the energy which the water has cannot be used to advantage, and *efficiency* is the ratio of the useful work done by the water to the total work done by it.

Efficiency

To find the amount of useful work done when a pump lifts or forces water to a higher level, multiply the weight of the water in pounds by the height in feet through which it is raised.

Since friction must be taken into consideration, the total work done by the pump will be greater than the useful work. To find the amount of the total work, multiply the amount of the useful work by the reciprocal of the efficiency of the pump.

Automobiles and Printing

Samuel Slade and Louis Margolis, 1941

This account of workshop mathematics is another thoroughly practical manual, by a team which was also responsible for the often reprinted *Mathematics for Technical and Vocational Schools* (1922). The latter has been translated into Spanish, and in a revised form was reprinted as recently as 2002. The extracts given here show how both old and new technologies required mathematical knowledge and procedures for those who worked with them.

Samuel Slade and Louis Margolis (dates unknown), *Essentials of Shop Mathematics* (New York, 1941), pp. 62–63, 103–105, 123–133.

Automobile mechanics

Work and power

When we lift a weight of one pound a height of one foot, we do *one foot-pound* of work. If we exert a pull of 50 pounds in dragging a load a distance of 10 feet, we do 10 × 50 or 500 foot-pounds of work.

A *foot-pound* is the unit of work. It is the work done in raising one pound a height of one foot, or it is the pressure of one pound exerted over a distance of one foot in any direction.

A freight elevator lifting 2,000 pounds a height of 40 feet does 80,000 foot-pounds of work whether it takes 2 minutes or 4 minutes to do it. But an elevator that would lift the load in 2 minutes would have twice the power of one that would take 4 minutes to do the same amount of work.

Power is the rate at which work is done, and is measured in foot-pounds per minute. Thus, if an elevator did 80,000 foot-pounds of work in 2 minutes, its power would be $\frac{80,000}{2}$ or 40,000 foot-pounds per minute. If it took 4 minutes to do it, its power would be $\frac{80,000}{4}$ or 20,000 foot-pounds per minute.

The unit of power is the *horsepower* and is equal to 33,000 foot-pounds per minute.

Example

An engine does 120,000 foot-pounds of work in 3 minutes. Compute the horsepower.

Solution

Horsepower = $\frac{120,000}{3 \times 33,000}$ = 1.2.

Explanation

Since it takes 3 minutes to do 120,000 foot-pounds of work, in 1 minute it will do $\frac{1}{3}$ of 120,000 or 40,000 foot-pounds of work.

Since 33,000 foot-pounds = 1 H.P., we divide by 33,000.

Problems

1. A man carries a load of 50 pounds up a flight of stairs 10 feet high. Compute the work done in foot-pounds.

2. Compute the horsepower required for an elevator if it is to lift 2,400 pounds a height of 120 feet in $1\frac{3}{4}$ minutes.

3. Find the horsepower of an engine that pumps 30 cubic feet of water per minute from a depth of 320 feet. Water weighs 62.5 pounds per cubic foot.

4. What must be the horsepower of an engine to lift a girder weighing 7 tons a height of 40 feet in 5 minutes?

5. A man in pushing a wheelbarrow a distance of 450 feet exerts a pressure of 45 pounds. How much work does he do?

Printing

The point system

The standard unit of measurement used by printers is the *pica*. There are six picas to the inch. The pica is divided into twelve parts or points.

Therefore there are 6 × 12, or 72, points to the inch.

$$1 \text{ inch} = 6 \text{ picas}$$

$$1 \text{ inch} = 72 \text{ points}$$

$$1 \text{ pica} = 12 \text{ points}$$

The pica is actually slightly less than one-sixth of an inch. The difference, however, is so small that for all practical purposes, it may be disregarded. The pica measures 0.16608 of an inch. Reduce one-sixth of an inch to a five-place decimal fraction and compare it with the pica.

Measuring types

The size of type is stated as the number of points in the height of the body of the piece of type. Thus, a piece of 10-point type is a piece of type which measures 10 points; a pica type is a piece of type which measures 12 points, or $\frac{1}{6}$ inch. Knowing the size of the type and the length of the page, we can compute the number of lines in the page or the length of the page required for a given number of lines.

Leaded type

⌞...⌟ In order to separate the lines and make the page easier to read, it is customary to insert strips of type metal between the lines. These are called *leads*. They are usually two points thick, though one-point leads and leads of other sizes are also used. This of course reduces the number of lines of type that can be printed on the page; or, if the same number of lines of type is required, the length of the page must be increased.

Measuring composed type

Another unit of measurement used in printing is the *em*. The em is a square of the type size used; that is, the width is the same as the height. For example, a 10-point em is a square 10 points on a side. The number of ems in a line depends on the size of the type.

The area of the printed page is expressed in *ems*. This is computed by multiplying the number of ems in the width of the page by the number of ems in the length of the page.

Space required

A common method for computing the amount of space or the number of pages of a stated size required for setting a given amount of copy is to use an average figure of 3 ems per word. The number of words in the copy is readily estimated by counting the number of words on one page of the typewritten or handwritten copy and multiplying by the number of pages.

Example 1

A manuscript consists of 42 typewritten pages and the average number of words on a page is 280. How many ems will the copy make?

Solution

$$42 \times 280 = 11{,}760 \text{ words}$$

$$11{,}760 \times 3 = 35{,}280 \text{ ems.}$$

Example 2

How many pages 23 picas by 40 picas will be required to set the job in Example 1 if 10-point type solid (without leads) is used?

Solution

$$\frac{23 \times 12}{10} = 27.6 \text{ ems to a line}$$

$$\frac{40 \times 12}{10} = 48 \text{ lines to a page}$$

$$48 \times 27.6 = 1{,}324.8 \text{ ems to a page}$$

$$35{,}280 \div 1{,}324 = 26.6$$

27 pages will be required.

❧ 9 ❧

"The Speedier Expedition of Their Learning": Thoughts on Teaching and Learning Mathematics

"ARITHMETIC IS A VERY DULL STUDY TO CHILDREN, AND IF THE ROD AND THE slate hang side by side, it cannot fail to be a disagreeable one," wrote Mrs. Lovechild in 1785. This chapter showcases a few of the many ways that mathematics teaching has been thought about and practiced over the years, and some of the ways different writers have tried to avoid making mathematics "dull and disagreeable."

We will see both examples of practice and more reflective discussions: Humfrey Baker's list of the mathematical subjects he taught to the "servants and children" at his London school, or the agonies of just how to present Euclid for the modern learner (with algebra? without axioms?) The dialogue format we have met in Chapters 2 and 4 receives more than one airing, continuing into the twentieth century.

We also pick up the story of arithmetic teaching, which in Chapter 2 we took no further than the early twentieth century. From the 1960s this subject became a divisive one, and the "New Math" appears here as, perhaps, the natural development of earlier concerns to teach methods rather than rote-learned rules. Chapter 8 has shown some examples of how that tension was played out in the mathematical workplace.

The most attractive theme of this chapter, though, and perhaps the most important, is that mathematics can and should be fun. The "mathematical toys" of the eighteenth century and the "game of logic" of the nineteenth receive a wonderful modern echo in one of my favorite extracts, the last in this chapter, on the mathematics of "turtle fun."

"To Have Their Children or Servants Instructed"

Humfrey Baker, 1590

We met Humfrey Baker in Chapter 1, as the author of the arithmetic textbook *The Welspring of Sciences*. Here we see a single-page advertisement for his mathematical school in London, showing in detail the "arts and faculties" which were taught there. Few hints are given of his teaching methods, but he has in common with later writers on mathematics pedagogy that he begins from a detailed understanding of how mathematics can be conceived and subdivided.

Humfrey Baker (fl. 1557–1574), *Such as are desirous* ⌐...⌐ ([London], c. 1590), single page.

Such as are desirous, either themselves to learn, or to have their children or servants instructed in any of these Arts and Faculties hereunder named, it may please them to repair unto the house of *Humfrey Baker*, dwelling on the North side of the *Royal Exchange*, next adjoining to the sign of the ship, where they shall find the professors of the said Arts, etc., ready to do their diligent endeavours for a reasonable consideration. Also, if any be minded to have their children boarded at the said house, for the speedier expedition of their learning, they shall be well and resonably used, to their contentation.

The Arts and Faculties to be taught, are these.

Arithmetic vulgar, namely Numeration, Addition, Subtraction, Multiplication, Division, Progression, in Whole numbers, ⌐and⌐ Fractions both Arithmetical and Astronomical. The application of ⌐this⌐ in operation, that is to say: The rule of 4 numbers proportional; The rule of 5 numbers or more; The rule of gain and loss with time; The rules of barter° both simple and compound; The rules of fellowship without time, and with time; The rules of interest both simple and compound; The rule of Alligation; The Virgins' rule; The rule of suppositions, commonly called false positions. ⌐All⌐ in Whole numbers, ⌐and⌐ Fractions. With the extraction of all kinds of roots of numbers, ⌐and the⌐ Numeration, Addition, Subtraction, Multiplication and Division of Proportions. The like in Surd and Cossic numbers,° with the application of them in the rule of Equation‚s‚, for Equations Arithmetical and Geometrical, and also the second quantities° of Algebra.

How to measure Lands, Woods, and all other platforms whatsoever, and to reduce the same platforms into greater or lesser shape, at your own pleasure; ‚and‚ Timber, Stone, and all kind of Solid bodies whatsoever.

The principles of Geometry, to be applied to the aid of all Mechanical workmen.

The use of the Quadrant, Geometrical square, and Baculus Jacob.°

The composition of the Astronomer‚'s‚ staff, Astrolabe, and Ruler of Ptolemy,° with their uses.

The calculation of the Tables of Sines, and Chords, with their proper use.

Also the perfect order how to keep any accounts by Debitor and Creditor, after a more plain manner than hath heretofore been usually taught by any man within this City. Likewise how to rectifie and make perfect any difficult or intricate account, depending in variance between two or more partners, and thereby to show which of them shall be indebted the one to the other.

Notes

rules of barter: rules that determine how much of one kind of goods should fairly be exchanged for a given amount of another if the price of each is known; the rules of fellowship determine how the profits should be shared from an enterprise where several people have contributed different amounts initially; the rule of alligation determines the price of a mixture or "mixed lot" of goods if the prices of the ingredients or components are known; the Virgin's rule is a mystery; while the rule of false position is in effect a strategy for solving linear equations by applying a correction to an initial guess.

Cossic numbers: unknown quantities or the multiples, squares, cubes, etc., of them, expressed using the sixteenth-century system of "cossic" notation, in effect an early form of algebraic notation.

second quantities: Baker may well mean areas, "first quantities" being lengths.

Quadrant, Geometrical square, Baculus Jacob: instruments used in surveying.

Astronomer's staff, Astrolabe, and Ruler of Ptolemy: astronomical instruments

---

Euclid with Algebra

Isaac Barrow, 1660

Isaac Barrow was Isaac Newton's predecessor as Lucasian Professor of Mathematics at Cambridge; later he left mathematics to concentrate on theology, and for many years he was remembered chiefly for his sermons. Also, for a time, a professor of Greek at Cambridge, he produced important editions of Euclid in both Latin and English. Here he introduces his Euclid and explains his aims in using the new algebraic notation wherever possible as a help to "the studious."

Isaac Barrow (1630–1677), *Euclide's Elements; The whole Fifteen Books compendiously Demonstrated By Mr Isaac Barrow Fellow of Trinity Colledge in Cambridge. And Translated out of the Latin.* (London, 1660), $\therefore 2^{\rm r}$–$\therefore 3^{\rm v}$.

In order to the Reader's satisfaction concerning the Book put into his hand, I am to advertise him of some few things, and that according to the nature of the Work, briefly, as followeth. My Undertaking aimed principally at two Ends: the first of which was to conjoin the greatest Compendiousness of Demonstration with as much Perspicuity as the quality of the subject would admit, that so the Volume might bear no bigger bulk than would render it conveniently portable. Which I have so far attained, that though possibly some other person might with greater curiosity, yet (I presume) none could with more conciseness have demonstrated most propositions, especially since I have altered nothing in the number and order of the Propositions, nor taken the liberty to leave out any one of Euclid's as less necessary, or to reduce certain of the easiest into the Class of Axioms.

❧

But I had a different Purpose from the beginning; not to compose Elements of Geometry any-wise at my discretion, but to demonstrate Euclid himself, and all of him, and that with all possible brevity. For as for Four of his Books, the Seventh, Eighth, Ninth and Tenth, although they do not so nearly pertain to the Elements of Plane and Solid geometry, as the Six First and the Two subsequent, yet no man that has arrived to any measure of skill in Geometry is ignorant how exceedingly useful they are in Geometrical matters, as well in regard of the very near alliance between Arithmetic and geometry, as for the knowledge of Commensurable and Incommensurable Magnitudes which is highly important to the understanding both of Plane and Solid Figures. And the noble Theory of the Five Regular Bodies, contained in the Three Last Books, could not be omitted without prejudice and injury; since our Author of these Elements, being a follower of Plato's School, is reported to have compiled the whole System only in reference to that Contemplation ⌊...⌋.

Moreover, I was easily induced to believe, that it would be acceptable to all Lovers of these Sciences to have the Entire work of Euclid by them, as it is usually cited and recommended by all men. Wherefore I determined to leave out no Book or Proposition ⸤...⸥. Upon the same account also I purposed to use generally no other than Euclid's own Demonstrations, contracted into a more succinct form, saving perchance in the Second and Thirteenth, and sparingly in the Seventh, Eighth, and Ninth Books, where it seemed convenient to vary something from him. So that it may be reasonably hoped that in this Particular our own Design and the Wishes of the Studious are in some manner satisfied.

The other End aimed at was in favour of Their desires who more affect Symbolical then Verbal Demonstrations. In which kind, seeing most of our own Nation are accustomed to the Notes of Mr. Oughtred,° I esteemed it more convenient to make use of them principally throughout. For no man hitherto that I know of, saving only Peter Herigon,° has attempted to set forth and interpret Euclid according to this way. The Method of which most learned Person, though in many other respects very excellent, and exactly accommodated to his peculiar purpose, seemed to me notwithstanding doubly defective. First, in that, whereas of several Propositions brought to the proving of some one Theorem or Problem the Latter does not always depend on the Former, yet when they do cohere one with another, and when not, cannot readily enough be known, either from their order or any other way; whence it not seldom comes to pass, that through the want of Conjunctions and Adjectives, *Ergo, rursus,*° etc., there arises difficulty and occasion of doubting, especially to such as are but little versed therein.

And in the next place, it oftentimes falls out that the said Method cannot avoid superfluous repetitions, whereby the Demonstrations become sometimes prolix, and sometimes perplexed and intricate. All which Inconveniences are easily remedied in our Way by the intermingling of Words and Signs at discretion. And thus much may suffice to be premised concerning the Intent and Method of this Compendium. I shall not allege in favour of myself the scantness of time allotted to this Work, nor the avocations of affairs, nor the scarcity of Helps to this sort of Studies amongst us (as I might not untruly) out of fear lest my Performance should not throughly please every body. But I wholly submit to the fair censure and Judgement of the Ingenuous Reader, what I have undertaken for the advantage of his Studies; to be approved, if he find it serviceable thereunto; or, if otherwise, rejected.

Notes

Mr. Oughtred: William Oughtred (1574–1660), an important English writer on algebra.
Peter Herigon: Pierre Hérigone, pseudonym of Baron Clément Cyriaque de Mangin (1580–1643). His translation of Euclid appeared posthumously in 1644.
Ergo, rursus: therefore, on the other hand.

The Idea of Velocity

Leonhard Euler, 1760

Leonhard Euler wrote his wildly popular *Letters to a German Princess* to the "young and sensible female," the princess of Anhalt-Dessau, at a rate of one or two each week, between 1760 and 1762. They ranged across the physical sciences and philosophy (including the intriguing "Electrification of men and animals") and covered a great deal of applied mathematics, always expressed using geometry rather than algebra. Euler's lucid but unpatronizing exposition makes the collected letters still one of the best introductions to the science of their period. Their English translator enthused about their potential to contribute to female education:

> The time, I trust, is at hand, when the Letters of Euler, or some such book, will be daily on the breakfasting table, in the parlour of every female academy in the kingdom; and when a young woman, while learning the useful arts of pastry and plain-work [sewing], may likewise be acquainting herself with the phases of the moon, and the flux and reflux of the tides.

Here we see Euler, near the beginning of the series of letters, introducing one of the fundamental concepts of applied mathematics: velocity.

Leonhard Euler (1707–1783), trans. Henry Hunter, *Letters of Euler to a German Princess, on different subjects in physics and philosophy* (London, 1795), vol. 1, pp. 6–9 (letter 2, dated 22 April 1760).

I proceed to unfold the idea of velocity, which is a particular species of extension, and susceptible of increase and of diminution. When any substance is transported, that is, when it passes from one place to another, we ascribe to it a velocity. Let two persons, the one on horseback, the other on foot, proceed from Berlin to Magdeburg; we have, in both cases, the idea of a certain velocity, but it will be immediately affirmed that the velocity of the former exceeds that of the latter. The question, then, is: Wherein consists the difference which we observe between these several degrees of velocity? The road is the same to him who rides, and to him who walks: but the difference evidently lies in the time which each employs in performing the same course. The velocity of the horseman is the greater of the two, as he employs less time on the road from Berlin to Magdeburg; and the velocity of the other is less, because he employs more time in travelling the same distance. Hence it is clear, that in order to form an accurate idea of velocity, we must attend at once to two kinds of quantity: namely to the length of the road, and to the time employed.

A body, therefore, which, in the same time, passes through double the space which another body does, has double its velocity; if, in the same time, it passes through thrice the distance, it is said to have thrice the velocity, and so on. We shall comprehend, then, the velocity of a body, when we are informed of the space through which it passes in a certain quantity of time. In order to know the velocity of my pace, when I walk to Lytzow,° I have observed that I make 120 steps in a minute, and one of my steps is equal to two feet and a half. My velocity, then, is such as to carry me 300 feet in a minute, and a space sixty times greater, or 18,000 feet, in an hour, which, however does not amount to a ,German, mile, for this, being 24,000 feet, would require an hour and 20 minutes. Were I, therefore, to walk from hence to Magdeburg, it would take exactly 24 hours. This conveys an accurate idea of the velocity with which I am able to walk.

Now it is easy to comprehend what is meant by a greater or less velocity. For if a courier were to go from hence to Magdeburg in 12 hours, his velocity would be the double of mine; if he went in eight hours, his velocity would be triple. We remark a very great difference in the degrees of velocity. The tortoise furnishes an example of a velocity extremely small. If she advances only one foot in a minute, her velocity is 300 times less than mine, for I advance 300 feet in the same time. We are likewise acquainted with velocities much greater. That of the wind admits of great variation. A moderate wind goes at the rate of 10 feet in a second, or 600 feet in a minute; its velocity therefore is the double of mine. A wind that runs 20 feet in a second is extremely violent, though its velocity is only 10 times greater than mine, and would take two hours and twenty-four minutes to blow from hence to Magdeburg.

The velocity of sound comes next, which moves 1000 feet° in a second, and 60,000 in a minute. This velocity, therefore, is 200 times greater than that of my pace; and were a cannon to be fired at Magdeburg, if the report could be heard at Berlin, it would arrive there in seven minutes. A cannon ball moves with nearly the same velocity; but when the piece is loaded to the utmost, the ball is supposed capable of flying 2,000 feet in a second, or 120,000 in a minute. This velocity appears prodigious, though it is only 400 times greater than that of my pace in walking to Lytzow; it is, at the same time, the greatest velocity known upon earth.

But there are in the heavens velocities far greater, though their motion appears to be extremely deliberate. You know that the earth turns round on its axis in 24 hours: every point of its surface, then, under the equator, moves ⸤24,840 English miles⸥, in 24 hours, while I am able to get though only ⸤about 83⸥ miles. Its velocity is, accordingly, 300 times greater than mine, and less, notwithstanding, than the greatest possible velocity of a cannon ball. The earth performs its revolution round the sun in the space of a year, proceeding at a rate of ⸤589,950 English miles⸥ in 24 hours. Its velocity, therefore, is 18 times more rapid than that of a cannon ball. The greatest velocity of which we have any knowledge is, undoubtedly, that of light, which moves ⸤9,200,000⸥ of miles every minute, and exceeds the velocity of a cannon ball, 400,000 times.

Notes

Lytzow: A village about a league from Berlin. (Translator's note.)

1000 feet: The velocity of sound is generally computed at 1,142 feet each second but varies with the elasticity and density of the air. The earth travels in her orbit 1,612,000 miles in the space of 24 hours and, therefore, with a velocity more than 50 times greater than that of a cannon ball. Light moves about 13 millions of miles every minute. (Translator's note.)

Mathematical Toys

"Mrs Lovechild," 1785

This unfortunately-named pedagogue produced a whole series of books aimed to help mothers teach their children; the emphasis was on making learning fun, and "Lovechild" described games and tricks as well as more orthodox teaching aids.

The selections given below show something of her philosophy of teaching as well as some of the mathematical games she considered suitable to engage the curiosity of small children.

"Mrs Lovechild" (Eleanor Fenn, 1743–1813), *The Art of Teaching in Sport; Designed as a Prelude to a set of Toys, for enabling Ladies to Instill the Rudiments of Spelling, Reading, Grammar, and Arithmetic, under the Idea of Amusement* (London, 1785), pp. 51–54, 60–62.

Arithmetic is a very dull study to children, and if the rod and the slate hang side by side, it cannot fail to be a disagreeable one.

There are amusing books, calculated to excite application in children in learning to read; but for figures, what near prospect of pleasure appears as an incentive?

A boy is required to learn accounts; he drudges in obedience to his parents, gets with difficulty through the first rules of Arithmetic, and contracts an aversion to figures for life.

Authority may place a child in the path of learning, but pleasure only can entice him on; let us therefore endeavour to strew the entrance with flowers, which may induce him to proceed with alacrity.

The Box ˌof mathematical toys, must be held sacred: the little people must not be allowed to touch it, nor to look in the book which contains the arcana.

The Box is to be produced occasionally, *as a favour*, and some of the sports indulged to the children, according to their progress. Ladies would do well to procure great abundance and variety of ˌwoodˌcuts, selected with care. The present set could then be distributed gradually, and replaced; thus the charm of novelty would long remain, and occasions of much instruction be introduced at a small expense.

Tricks with Cards.

I.

Turn the cards. For instance, a five. Then an eight. A young child adds the numbers: 8 and 5 are 13. Or an older one subtracts the smaller number:

5 from 8, and there remain 3. Or a still older multiplies the two numbers together: five times 8 are 40.

II.

Take nine cards, *viz.* Ace, ‚two, three‚, etc. Place them so as to make 15 eight ways.

$$
\begin{array}{ccc}
4 & 3 & 8 \\
9 & 5 & 1 \\
2 & 7 & 6 \\
\end{array}
$$

A lady may not choose to take the trouble of discovering how they are to be placed—though she will contentedly drudge at 2 and 2, for the benefit of her children—so the order is shown. A child must be pretty well versed in figures before he will be able to *discover* how they are to be placed.

III.

The pack of cards ‚without court cards, being taken,‚ the whole amount of the pips of a pack (not counting the tenth cards) is 180. One player takes a card. The other is to discover what that card is, by missing the number of it. Thus: The player is requested to take a card below ten; suppose ‚she‚ takes a four. I then miss four from 180, and have only to run the pack over again, to see *which* four is absent. Or, if you count the cards as they pass in review before you, casting out ‚tens . . . ‚, till you see what is wanted of the last ten. This requires a readiness in addition to do it well. For a young child, take cards whose pips amount to 100. For a very young one, ‚take cards whose pips amount to‚ 20 only, and that number composed of small cards.

The Merchant, or Commerce.

A merchant sold beans; he was of so suspicious a disposition that he apprehended every person meant to impose on him; he was never satisfied with telling his money or beans once or twice, but counted them several times, and in every possible manner. If he had twenty beans he would first count them thus: two and two are four, four and two are six, and so on by two at a time to twenty. Then he counted 3 and 3 are 6, and 3 are 9, and 3 are 12, and 3 are 15, and 3 are 18, and 2 are 20. Then 4 and 4 are 8, and so on to twenty. Then 5 and 5 are 10, and so on to twenty. Then 6 and 6, etc.

This sport should be enlivened by secreting a counter or bean (occasionally), ⸤for⸥ the child to miss it.

A Mother Explains Comets

Catherine Vale Whitwell, 1823

Comparable in some ways to *Newton for the Ladies* (Chapter 5), this account of astronomical theories in dialogue form illustrates one author's concern that mathematical subjects be made accessible even to the very young, and to girls. Catherine Vale Whitwell, also the author of a "scheme of simplicity and economy" addressed to mothers, is ambitious in the complexity of the astronomical information she relates (much of it taken from Charles Hutton's mathematical dictionary of 1795–1796), although she stops short of giving a quantitative or geometric account of comets' orbital motions.

Catherine Vale Whitwell, *An Astronomical Catechism; or, Dialogues between a mother and her daughter* (London, 1823), pp. 124–129.

Are not comets important bodies belonging to our system? Will you kindly inform me, how I may distinguish them?

Comets are solid opaque bodies, of different magnitudes, like the planets, but are distinguished by fiery tails, or long beards of transparent hair.

From which side does the tail issue?

From that side of them which is furthest from the sun.

Has the tail the same degree of lustre throughout?

No: the lustre is the most considerable near the body of the comet, and becomes less and less, the further it is removed thence.

Is there not a popular division of comets into three kinds?

Yes; there is such a division, but it rather relates to the several circum-stances of the same comet, than to the phenomena of several: the divisions are the bearded, the tailed, and the hairy comet.

Will you explain this remark by an illustration?

When a comet is eastward of the sun, and moves from it, it is said to be bearded, because the light precedes it, in the manner of a beard.

What is it termed when the comet is westward of the sun, and sets after it?

It is said to be tailed, because the transparent hairlike appendage follows it in the manner of a tail.

When the comet and the sun are diametrically opposite, the earth being between them, what is the term employed?

Then the train is hid behind the body of the comet, except a little which appears round it, in the form of a border or tuft of hair, or coma; hence it is called hairy, and from hence the name comet is derived.

Is the body as well as the tail of the comet subject to apparent changes?

Yes; and these Sir Isaac Newton ascribes to changes in the atmosphere of the comet. His opinion was confirmed by observations on the comet of 1774.

At what periods have very interesting comets made their appearance?

In Dr. Rees' *Cyclopedia* you will find the elements of ninety-seven comets, and the names of the authors who have calculated their orbits. But a few words respecting the comets of 1680, 1661, and 1664, will answer my purpose.

As the elements of so large a body of comets are given, they have, I suppose, in common with other heavenly bodies, claimed the attention of astronomers in every age?

Not so; and nothing is more to be regretted by posterity, than that the ancients considered them as portentous signals of approaching calamity, instead of regarding them in a philosophical point of view.

What, then, were they altogether neglected by ancient astronomers?

They were, at most, considered only as meteors, or sublunary vapors floating in the atmosphere.

Whose attention did they first excite, and by whom were they first regarded as bodies of more importance?

Seneca, who was born six years before Christ, paid considerable atten-tion to them, in consequence of the appearance of two in his own time. But his observations remain unrecorded. Who then was the first, who described with any degree of accuracy the path of a comet?

Nicephorus Gregorius,° an astronomer and historian of Constantinople. Deeply impressed with the neglect of his predecessors, he composed that stupendous effort of human industry, "The Table of the Elements of Comets."

O mamma, with what multiplied specimens of energy, and with what vast achievements of industry, do you furnish me! Surely I ought to catch one spark of ambition from the fire burning on that altar. With a reputation ever gathering, shall the name of such a man descend to posterity. But may I ask, what was the appearance of the comet of 1680?

Sturmius° tells us that when he viewed it with a telescope, it appeared like a coal dimly glowing, or a rude mass of matter, illuminated with a dusky fumed light, less sensible at the extremes than in the centre.

What was the appearance of the comet of 1661?

Hevelius observed, that its body was of a yellowish colour, very bright and conspicuous, but without glittering light. In the middle was a dense ruddy nucleus, almost equal to Jupiter, encompassed with a much fainter, thinner matter.

But do these instances evince that the body of the comet is subject to apparent changes?

No, they do not; therefore I will enter a little more into the detail. On the 5th of February, the head of the last-mentioned comet was somewhat bigger, and brighter, than it had before been; it was of a gold colour, but its light was more dusky than the rest of the stars, whilst the nucleus appeared divided into several parts. February the 6th, the disk was lessened, the nuclei still existed, though less than before; February the 8th, its body was round, and represented a very lucid little star; the nuclei were still encompassed with another kind of matter. February the 10th, the head was somewhat more obscure, and the nuclei more confused, but brighter at the top than at the bottom. February the 13th, the head was much diminished, both in magnitude and brightness. March the 2nd, its roundness was a little impaired, its edges lacerated, etc. March the 28th, it was very pale and exceedingly thin; its matter was much dispersed, and no distinct nucleus appeared.

What is the description given of the comet?

Weigelius° says that he saw the comet, the moon, and a little cloud illumined by the sun at the same time; and he observed that the moon, through a telescope, appeared of a continued luminous surface, but the comet very different, being exactly like a little cloud in the horizon illumined by the sun.

I have heard various conjectures respecting the cause of the tails of the comets; will you have the goodness to mention to me the most plausible opinion?

It is that formed by Hevelius,° who supposes that the thinnest parts of the atmosphere of a comet are rarefied by the force of the heat, and driven from the forepart and from each side of the comet, towards the parts turned from the sun.

What is the greatest distance from the sun at which a comet can be situated, to be visible to us on the earth?

About three times the distance of the sun from the earth.

At what distance from the sun was the comet of 1680, at the period of its greatest heat?

About four-sevenths of the earth's distance from the sun, or forty-six millions, two hundred and eighty-five thousand miles, nearly.

When a comet has been once seen, can the period of its return be foretold?

Not with any degree of certainty, though it is possible that the periods of three of them may have been ascertained. The first of these appeared in the years 1531, 1607, and 1682, and is expected to return every 75th year: there is also another, which by its return in 1758 gratified astronomers, as its appearance corresponded with the prediction of Dr. Halley.°

When did the second appear?

In the years 1522, and 1662; and it was expected that it would again make its appearance in 1789, but in this the astronomers of the present day have been disappointed.

Notes

Nicephorus Gregorius: or *Gregoras* (c. 1295–1360), remembered for his historical writings.

Sturmius: Johann Christophorus Sturm (1635–1703), German astronomer and mathematician.

Weigelius: Erhard Weigel (1625–1699), German mathematician, astronomer, and philosopher.

Hevelius: Johannes Hevelius (1611–1687), politician and astronomer, of Danzig.

Dr. Halley: Edmond Halley (1656–1742), who had indeed predicted that the comet now named after him would reappear, although the details were worked out by others.

"Geometry without Axioms"

Thomas Perronet Thompson, 1833

Something of an oddity among nineteenth-century Euclids, of which there were not surprisingly several, Thompson's attempted, as the title page explained, "to get rid of Axioms and Postulates; and particluarly to establish the Theory of Parallel Lines without recourse to any principle not grounded on previous demonstration." It seems to have been motived at least in part by unease about the status of the "parallel postulate," but one cannot help feeling that in rewriting Euclid without axioms a baby has been lost as well as some bathwater. By the time we reach proposition I.47, however, things look more or less as in most Euclids.

Thomas Perronet Thompson (1783–1869) is mainly remembered as an army officer and politician; he had been governor of Sierra Leone by the time his Euclid appeared and would later distinguish himself as a political writer and radical MP.

Thomas Perronet Thompson, *Geometry without Axioms, or The First Book of Euclid's Elements. With Alterations and Familiar Notes; and an Intercalary Book in which the straight line and plane are derived from properties of the Sphere.* (London, 4th edition, 1833), pp. v–ix, 1–2, 119–120.

In the preceding Editions endeavour had been made to get rid of Axioms, and particularly to establish the Theory of Parallel Lines without recourse to any principle not grounded on previous demonstration.

On showing the results to some of the leading mathematicians at Cambridge, they replied by the remark that they had always felt something to be more urgently wanted for the emendation of Geometry—which was, information on the nature and construction of the straight line and plane.

It had been stated, about the time when the circumstances were engrossing the attention of the public, that Napoleon on his voyage from Egypt amused himself and staff with circular geometry. What circular geometry might be, could only be collected from the tradition that the problem given by the future Emperor was "to divide the circumference of a circle into four equal parts by means of circles only." But this sufficed to indicate that the idea which had passed through the mind of that eminent practical geometer was that in the properties of the circle, or still more probably of the sphere, might be discovered the elements of geometrical organization.

The author had in consequence been led at different times to attempt the collecting of the conditions, under which figures of various kinds may be turned about certain points and be what may be termed introgyrant, or turn upon their own ground without change of place. And on receipt of the remark mentioned above, this track was pursued with renewed vigour, and the results are presented in the sequel.

These results, together with the proofs offered of what have usually been termed the Axioms and Postulates, have been formed into an Intercalary Book, with a view to facilitate their postponement in the case of students beginning geometry for the first time. And for the further convenience of this class, a Recapitulation of the principal contents of the Intercalary Book has been given ⌞...⌟, thereby placing the beginner in the same situation as by the ordinary proceeding. The best time for commencing the Intercalary Book would probably be after having gone once through as much of the Elements of Euclid as is usually read; not, of course, that the contents are in any degree dependent upon what follows, but to take advantage of the habits of reasoning that may be thus acquired, before attempting what must be characterized as, in some parts, at least equal in complexity to anything in the succeeding Books.

If this process is objected to as irregular, it may be a great irregularity that nature should not have framed the elements of geometry so as to present a ⌞harmonious⌟ whole with the easiest parts always foremost and the Planes in the Eleventh Book. But if she has not, or till somebody can establish that she has, there seems to be no cause why bad reasoning should be admitted for the sake of a conventional arrangement. If the sphere is the simplest of figures and the properties of all others are derivable from it, it is more reasonable to be thankful for the knowledge than to quarrel with the dispensation. ⌞...⌟

In labouring to get rid of Axioms, the object has been to assail the belief in the existence of such things as self-evident truths. Nothing is self-evident, except perhaps an identical proposition. There may be things of which the evidence is continually before the senses; but these are not self-evident, but proved by the continual evidence of the senses. There may be things whose connexion with other things is so constantly impressed upon us by experience that few people ever think of inquiring into the cause, but for that very reason there is often considerable difficulty in clearly explaining the cause, and among this class of things the admirers of axioms have found their greatest crop of self-evident truths. In arguments on the general affairs of life, the place where every man is most to be suspected is in what he starts from as "what nobody can deny." It was therefore of evil example that science of any kind should be supposed to be founded on axioms, and it is no answer to say that in a particular case they were true. The Second Book of Euclid would be true, if the First existed only in the shape of the heads of the Propositions under the title of Axioms, but this would make a most lame and imperfect specimen of reasoning.

The Game of Logic

Lewis Carroll, 1887

It would be a strange anthology of popular mathematics that passed by the nineteenth century without mentioning Lewis Carroll, as Charles Lutwidge Dodgson is better known. Few readers will need to be introduced to *Alice's Adventures in Wonderland* or *The Hunting of the Snark*, but Dodgson wrote much else, including this attractive book, *The Game of Logic*, which I hope may be less familiar.

The book is introduced by the following splendid line: "There foam'd rebellious Logic, gagg'd and bound"—which turns out (Carroll does not tell us) to be from Alexander Pope's mock epic *The Dunciad* (1728/1741); logic, there, was "gagg'd and bound," alongside science, wit, and rhetoric, beneath the throne of "dullness" or stupidity.

Lewis Carroll, *The Game of Logic* (London, 1887), pp. 1–6.

Preface

This Game requires nine Counters—four of one colour and five of another, say four red and five grey.

Besides the nine Counters, it also requires one Player, AT LEAST. I am not aware of any Game that can be played with LESS than this number, while there are several that require MORE: take Cricket, for instance, which requires twenty-two. How much easier it is, when you want to play a Game, to find ONE Player than twenty-two. At the same time, though one Player is enough, a good deal more amusement may be got by two working at it together, and correcting each other's mistakes.

A second advantage possessed by this Game is that, besides being an endless source of amusement (the number of arguments that may be worked by it being infinite), it will give the Players a little instruction as well. But is there any great harm in THAT, so long as you get plenty of amusement?

Propositions.

> "Some new Cakes are nice."
> "No new Cakes are nice."
> "All new cakes are nice."

There are three "PROPOSITIONS" for you—the only three kinds we are going to use in this Game; and the first thing to be done is to learn how to express them on the Board.

Let us begin with

> "Some new Cakes are nice."

But before doing so, a remark has to be made—one that is rather important, and by no means easy to understand all in a moment; so please to read this VERY carefully.

The world contains many THINGS (such as "Buns," "Babies," "Beetles," "Battledores," etc.), and these Things possess many ATTRIBUTES (such as "baked," "beautiful," "black," "broken," etc.: in fact, whatever can be "attributed to," that is "said to belong to," any Thing, is an Attribute). Whenever we wish to mention a Thing, we use a SUBSTANTIVE; when we wish to mention an Attribute, we use an ADJECTIVE. People have asked the question "Can a Thing exist without any Attributes belonging to it?" It is a very puzzling question, and I'm not going to try to answer it; let us turn up our noses, and treat it with contemptuous silence, as if it really wasn't worth noticing. But, if they put it the other way, and ask "Can an Attribute exist without any Thing for it to belong to?", we may say at once "No: no more than a Baby could go a railway-journey with no one to take care of it!" You never saw "beautiful" floating about in the air, or littered about on the floor, without any Thing to BE beautiful, now did you?

And now what am I driving at, in all this long rigmarole? It is this. You may put "is" or "are" between names of two THINGS (for example, "some Pigs are fat Animals"), or between the names of two ATTRIBUTES (for example, "pink is light-red"), and in each case it will make good sense. But, if you put "is" or "are" between the name of a THING and the name of an ATTRIBUTE (for example, "some Pigs are pink"), you do NOT make good sense (for how can a Thing BE an Attribute?) unless you have an understanding with the person to whom you are speaking. And the simplest understanding would, I think, be this—that the Substantive shall be supposed to be repeated at the end of the sentence, so that the sentence, if written out in full, would be "some Pigs are pink (Pigs)." And now the word "are" makes quite good sense.

Thus, in order to make good sense of the Proposition "some new Cakes are nice," we must suppose it to be written out in full, in the form "some new Cakes are nice (Cakes)." Now this contains two "TERMS"— "new Cakes" being one of them, and "nice (Cakes)" the other. "New Cakes," being the one we are talking about, is called the "SUBJECT" of the Proposition, and "nice (Cakes)" the "PREDICATE." Also this Proposition is said to be a "PARTICULAR" one, since it does not speak of the WHOLE of its Subject, but only of a PART of it. The other two kinds are said to be "UNIVERSAL," because they speak of the WHOLE of their Subjects— the one denying niceness, and the other asserting it, of the WHOLE class of "new Cakes." Lastly, if you would like to have a definition of the word "PROPOSITION" itself, you may take this: "a sentence stating that some, or none, or all, of the Things belonging to a certain class, called its 'Subject', are also Things belonging to a certain other class, called its 'Predicate'."

You will find these seven words—PROPOSITION, ATTRIBUTE, TERM, SUBJECT, PREDICATE, PARTICULAR, UNIVERSAL— charmingly useful, if any friend should happen to ask if you have ever studied Logic. Mind you bring all seven words into your answer, and you friend will go away deeply impressed—"a sadder and a wiser man."

Now please to look at the smaller Diagram of the Board (Figure 9.1), and suppose it to be a cupboard, intended for all the Cakes in the world (it would have to be a good large one, of course). And let us suppose all the new ones to be put into the upper half (marked "x"), and all the rest (that is, the NOT-new ones) into the lower half (marked "x'"). Thus the lower

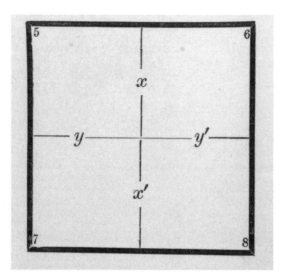

Figure 9.1. The playing board for the Game of Logic. ©Princeton University Library. Rare Books Division, Department of Rare Books and Special Collections, BC135 .D6.

half would contain ELDERLY Cakes, AGED Cakes, ANTE-DILUVIAN Cakes—if there are any: I haven't seen many, myself—and so on. Let us also suppose all the nice Cakes to be put into the left-hand half (marked "y"), and all the rest (that is, the not-nice ones) into the right-hand half (marked "y′"). At present, then, we must understand x to mean "new," x′ "not-new," y "nice," and y′ "not-nice."

And now what kind of Cakes would you expect to find in compartment No. 5?

It is part of the upper half, you see, so that, if it has any Cakes in it, they must be NEW; and it is part of the left-hand half, so that they must be NICE. Hence if there are any Cakes in this compartment, they must have the double "ATTRIBUTE" "new and nice;" or, if we use letters, they must be "x y."

Observe that the letters x, y are written on two of the edges of this compartment. This you will find a very convenient rule for knowing what Attributes belong to the Things in any compartment. Take No. 7, for instance. If there are any Cakes there, they must be "x′ y," that is, they must be "not-new and nice."

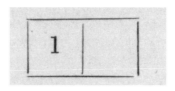

Figure 9.2. Fixing our attention on this upper half, suppose we found it marked like this. (Carroll, p. 6. ©Princeton University Library. Rare Books Division, Department of Rare Books and Special Collections, BC135 .D6.)

Now let us make another agreement—that a red counter in a compartment shall mean that it is "OCCUPIED," that is, that there are SOME Cakes in it. (The word "some," in Logic, means "one or more" so that a single Cake in a compartment would be quite enough reason for saying "there are SOME Cakes here"). Also let us agree that a grey counter in a compartment shall mean that it is "EMPTY," that is that there are NO Cakes in it. In the following Diagrams, I shall put "1" (meaning "one or more") where you are to put a RED counter, and "0" (meaning "none") where you are to put a GREY one.

As the Subject of our Proposition is to be "new Cakes," we are only concerned, at present, with the UPPER half of the cupboard, where all the Cakes have the attribute x, that is, "new."

Now, fixing our attention on this upper half, suppose we found it marked like this (see Figure 9.2), that is, with a red counter in No. 5. What would this tell us, with regard to the class of "new Cakes"?

Would it not tell us that there are SOME of them in the x y-compartment? That is, that some of them (besides having the Attribute x, which belongs to both compartments) have the Attribute y (that is, "nice"). This we might express by saying "some x-Cakes are y-(Cakes)," or, putting words instead of letters,

"Some new Cakes are nice (Cakes),"
or, in a shorter form,
"Some new Cakes are nice."

At last we have found out how to represent the first Proposition of this Section. If you have not CLEARLY understood all I have said, go no further, but read it over and over again, till you DO understand it. After that is once mastered, you will find all the rest quite easy.

Higher Mathematics for Women

Mrs. Henry Sidgwick, 1912

In 1911–1912 the British Board of Education assembled a volume of papers reporting on the teaching of mathematics for presentation to the International Congress of Mathematicians in Cambridge in 1912. The eighteenth paper in the series discussed "mathematics in the education of girls and women," and the three authors displayed a range of views on the subject. Eleanor Mildred Sidgwick (her husband was the philosopher Henry Sidgwick), in her section titled "Higher Mathematics for Women," was easily the most optimistic, in line with her role as principal of the all-female Newnham College, Cambridge, and a promoter of university education for women. Her warning against teaching and learning mathematics only "because it is useful for something else" retains its force in the twenty-first century.

Mrs. Henry Sidgwick (1845–1936), "Higher Mathematics for Women," in E. R. Gwatkin, Sara A. Burstall, and Mrs. Henry Sidgwick, "Mathematics in the Education of Girls and Women"; The Board of Education, *Special Reports on Educational Subjects: The Teaching of Mathematics in the United Kingdom*, no. 18 (London, 1912), pp. 26–27, 31.

I have been asked, in connection with the work of the International Commission on the Teaching of Mathematics, to write a paper on the above subject, and my attention has been specially invited to the question whether, for instance, the study of Mathematics by women should have a special bearing on the subsequent study of Economics, or of statistical inquiries into sociological questions, and therefore perhaps differ in kind from the Mathematics required by, e.g., a future engineer. I understand the expression Higher Mathematics to be used, not in the sense that mathematicians might give it, but merely to mean Mathematics as studied at the Universities by those who are at least to some extent specialising in the subject, as distinct from the Mathematics which form a part of general education at secondary schools.

It will perhaps conduce to clearness if I state at once that, for reasons set forth below, it does not appear to me necessary or advisable to have separate programmes of mathematical study for men and for women at the

Universities. If an inherent difference in mathematical ability between the two sexes were established it might seem at first sight to furnish a reason for a difference of programme. But experience has shown that some women have sufficient ability and sufficient liking for Mathematics to justify their making it their principal subject of study at the University, and this being so, the inquiry how their powers compare with those of men, either on the average, or as regards the highest degrees of mathematical ability, becomes irrelevant to the question before us. For the great differences in ability among the men who specialise in Mathematics afford a range amply wide enough to include all the women who could on any hypothesis be reasonably advised to do so.

In considering the nature of the demand for mathematical education on the part of women, I may take as fairly typical the present (1911–12) students of Newnham College at Cambridge. This is convenient, both because information about them is easily accessible to me, and because the extent to which specialisation is carried at Cambridge, and the consequent sharp differentiation between the different Triposes, makes classification comparatively easy. I should explain that, with the exception of a few graduates from other Universities, and a few other students taking special courses, the women students at Cambridge, whether at Girton or Newnham,° are reading for Honours in one or other of the Tripos examinations, i.e., examinations for degrees in Honours. Out of 213 students at Newnham in November 1911 were 30 reading Mathematics, 2 Engineering, 35 Classics, 3 Moral Sciences, 1 Law, 33 Natural Sciences (including Geography), 39 History, 61 Medieval and Modern Languages, 9 Economics. Over 14 per cent of the students therefore were working at Mathematics, and this proportion is somewhat less than has been usual in other years. Of the 30 reading Mathematics, 9 were in their first year, 13 in their second year, and 8 in their third year at the University. A certain number will not continue to work at Mathematics through their whole course, but after obtaining Honours in Part I of the Mathematical Tripos at the end of their first or second year will turn their attention to some other subject, e.g., Science, especially Practical Physics for the first or second part of the Natural Science Tripos, or Geography for the Diploma in Geography, or possibly Economics for the Economics Tripos. They might even take some subject unconnected with Mathematics, such as History or Languages. The proportion of students of Mathematics is usually much the same at Girton as at Newnham, but probably the proportion at Cambridge is in excess of that among women students at the Universities generally, as the prestige of Cambridge in Mathematics no doubt tends to attract mathematical students there.

The next question to be considered is what leads these women to make a special study of mathematics. It is almost always, I think, that they have liked the subject and succeeded fairly well in it during their school education, and have consequently been advised by their school teachers to go on with it at the University as a subject they are likely to do well in. If, as is frequently the case, they intend to become teachers, they think it is the subject they will like best to teach, and which they are most likely to teach well and to find profitable employment in. The estimate thus formed of their powers and tastes, either by the students themselves or their teachers, does not always prove correct. It happens occasionally that facility in manipulating algebraical formulae has been mistaken for mathematical grasp. It happens, occasionally, that taste for the subject diminishes as the difficulty increases. It happens, also, sometimes that the amount of previous preparation required for success in the Cambridge course, by any but persons of unusual ability, is under-estimated, and that a student who has come to the University with the intention of studying Mathematics, but insufficiently prepared, has to be advised to change her plans. It does not often happen that, on the other hand, a student who comes up intending to devote herself to some other subject, changes to Mathematics.

If I am right in the above diagnosis of the motives that lead women to the study of Mathematics at Cambridge, the subject is in almost all cases studied mainly for its own sake—not because it is useful as an adjunct or stepping-stone to something else. And educationally it is very important that this should be so. A subject which is studied, not for its own sake, but because it is useful for something else, is almost always degraded in the process, and loses much of its educational value, whether the ultimate object be merely to pass an examination or to acquire the minimum knowledge necessary for dealing with some other and different subject of study. It is, no doubt, quite possible to learn Mathematics after a fashion, and up to a certain point, purely as an instrument for some ulterior purpose. But probably for everyone, and certainly for anyone with any mathematical ability, the loss from such a method of study would be great. There would be loss in mental training, loss in knowledge, and loss in interest. A recent writer in the Engineering Supplement of "The Times" advocating a continued attention to Mathematics for its own sake on the part of engineers, said, "Its importance in education is to form deliberate judgment, to assign a measure to results, to disclose fallacies, to discourage narrowness, to reveal the unity of whatsoever things are true, and generally to exalt the mind." I should be inclined to add, to stimulate the imagination. The study of mathematics reveals to the eyes that can see whole vistas of knowledge, whole aspects of the universe, unsuspected by those without mathematical perception.

I would not, therefore, limit the opportunities of pursuing the study of Mathematics for its own sake in any way, either for women or for men. It must not of course, be forgotten that among those who study Mathematics at the Universities a few—one here and there—may have sufficient power and originality to carry the subject in some direction or other beyond the present limits of our knowledge. But quite apart from this, it is important to keep before everyone, and before women especially, the value of knowledge for its own sake. I say before women especially, because I think the education of women has suffered more than that of men from what I may call, for shortness, the commercial or utilitarian point of view. The old bad education so prevalent in the girls' schools and the private schoolrooms of 50 or 60 years ago—drawing-room accomplishments, needlework, a little arithmetic and something of general information—was a kind of degenerate technical education for domestic life. The standard was low because neither thoroughness nor a high standard were needed for the purposes aimed at, and the standard tended to get worse rather than better because the teachers, being taught in the same schools, had no better opportunities of learning than their pupils. As a general result extremely little intellectual training or cultivation was, as a rule, attained, and very little interest was developed in anything beyond trivialities. I fear a return towards a similar state of things when I hear it suggested that Science should be taught to girls and women as the handmaid of Cookery and Hygiene, mathematics as the handmaid of Economics, and Economics again studied because a knowledge of Economics may be useful in social work.

⌊· · ·⌋

To sum up, my conclusion is that in the first place it is highly desirable that women with mathematical ability, even if it be not of a very high order, should be encouraged to study the subject for its own sake, and not with limits prescribed by the utility of Mathematics for something else, for it is thus that they will get the utmost value out of it both in pleasure and in mental training. And I conclude in the second place that so far as it may be necessary to make arrangements for the study of mathematics merely as a stepping stone to other studies, there is no need to consider the case of women separately from that of men.

Note

Girton or Newnham: At this time these were the only two Cambridge colleges which admitted women, and they both admitted only women.

A New Aspect of Mathematical Method

George Pólya, 1945

George Pólya, a Hungarian emigré to the United States, worked in real and complex analysis, as well as other areas of mathematics, and also studied mathematical pedagogy. *How to Solve It*, his book on mathematical heuristics, strategies of problem solving, and how to teach and learn them, is said to have sold over a million copies in seventeen languages and is one of the classics of modern popular mathematics.

   While the dialogue form is one we have met before, in Catherine Whitwell's *Astronomical Catechism* and in *Newton for the Ladies* in Chapter 5, what Pólya intends to convey by it, concerning the learning of mathematics as a method of thought, has a distinctly different flavor from anything else in this book.

G. Pólya (1887–1985), *How to Solve It: A New Aspect of Mathematical Method* (Princeton, 1945, 1957; London, 1990), pp. 25–29. © 1945 Princeton University Press, 1973 renewed by Princeton University Press, 2004 Princeton University Press, Expanded Princeton Science Library. Reprinted by permission of Princeton University Press.

A problem to prove.

Two angles are in different planes but each side of one is parallel to the corresponding side of the other, and has also the same direction. Prove that such angles are equal.

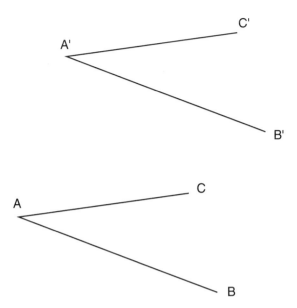

Figure 9.3. Two angles are in different planes, but each side of one is parallel to the corresponding side of the other and has also the same direction.

What we have to prove is a fundamental theorem of solid geometry. The problem may be proposed to students who are familiar with plane geometry and acquainted with those few facts of solid geometry which prepare the present theorem in Euclid's Elements. (The theorem that we have stated and are going to prove is the proposition 10 of Book XI of Euclid.) ⌊...⌋

"*What is the hypothesis?*"

"Two angles are in different planes. Each side of one is parallel to the corresponding side of the other, and has also the same direction."

"*What is the conclusion?*"

"The angles are equal."

"*Draw a figure. Introduce suitable notation.*"

The student draws the lines of Figure 9.3 and chooses, helped more or less by the teacher, the letters as in the figure.

"*What is the hypothesis? Say it, please, using your notation.*"

"A, B, C are not in the same plane as A', B', C'. And $AB \parallel A'B'$, $AC \parallel A'C'$. Also AB has the same direction as $A'B'$, and AC the same as $A'C'$."

"*What is the conclusion?*"

"$\angle B\,AC = \angle B'\,A'C'$."

"*Look at the conclusion! And try to think of a familiar theorem having the same or a similar conclusion.*"

"If two triangles are congruent, the corresponding angles are equal."

"Very good! Now *here is a theorem related to yours and proved before. Could you use it?*"

"I think so but I do not see yet quite how."

"*Should you introduce some auxiliary element in order to make its use possible?*"

❧

"Well, the theorem which you quoted so well is about triangles, about a pair of congruent triangles. Have you any triangles in your figure?"

"No. But I could introduce some. Let me join B to C, and B' to C'. Then there are two triangles, $\triangle ABC$, $\triangle A'B'C'$."

"Well done. But what are these triangles good for?"

"To prove the conclusion, $\angle B\,AC = \angle B'\,A'C'$."

"Good! If you wish to prove this, what kind of triangles do you need?"

"Congruent triangles. Yes, of course, I may choose B, C, B', C' so that

$$AB = A'B', \; AC = A'C'.$$

"Very good! Now, what do you wish to prove?"

"I wish to prove that the triangles are congruent,

$$\triangle ABC = \triangle A'B'C'.$$

If I could prove this, the conclusion $\angle B\,AC = \angle B'A'C'$ would follow immediately."

"Fine! You have a new aim, you aim at a new conclusion. *Look at the conclusion! And try to think of a familiar theorem having the same or a similar conclusion.*"

"Two triangles are congruent if—if the three sides of the one are equal respectively to the three sides of the other."

"Well done. You could have chosen a worse one. Now *here is a theorem related to yours and proved before. Could you use it?*"

"I could use it if I knew that $BC = B'C'$."

"That is right! Thus, what is your aim?"

"To prove that $BC = B'C'$."

"*Try to think of a familiar theorem having the same or a similar conclusion.*"

"Yes, I know a theorem finishing: '... then the two lines are equal.' But it does not fit in."

"*Should you introduce some auxiliary element in order to make its use possible?*"

⟜

"You see, how could you prove $BC = B'C'$ when there is no connection in the figure between BC and $B'C'$?"

...

"*Did you use the hypothesis? What is the hypothesis?*"

"We suppose that $AB \parallel A'B'$, $AC \parallel A'C'$. Yes, of course, I must use that."

"*Did you use the whole hypothesis? You say that $AB \parallel A'B'$. Is that all that you know about these lines?*"

"No; AB is also equal to $A'B'$, by construction. They are parallel and equal to each other. And so are AC and $A'C'$."

"Two parallel lines of equal length—it is an interesting configuration. *Have you seen it before?*"

"Of course! Yes! Parallelogram! Let me join A to A', B to B', and C to C'."

"The idea is not so bad. How many parallelograms have you now in your figure?"

"Two. No, three. No, two. I mean, there are two of which you can prove immediately that they are parallelograms. There is a third which seems to be a parallelogram; I hope I can prove that it is one. And then the proof will be finished!"

We could have gathered from his foregoing answers that the student is intelligent. But after this last remark of his, there is no doubt.

This student is able to guess a mathematical result and to distinguish clearly between proof and guess. He knows also that guesses can be more or less plausible. Really, he did profit something from his mathematics classes; he has some real experience in solving problems, he can conceive and exploit a good idea.

New Math for Parents

Evelyn Sharp, 1966

Evelyn Sharp's book aimed "to explain to parents something of the tremendous changes taking place in ...school mathematics" during the 1960s. She admitted that "there must be many mothers and fathers who are bewildered by the new approach to the subject" and gave a readable and comprehensible approach to mathematics along New Math lines. Here we have her outline of the motivations for the new methods and her discussion of the basic laws of arithmetic, forming a remarkable contrast with the older discussions of arithmetic in Chapter 2 of this book.

Evelyn Sharp, edited by L. C. Pascoe, *A Parent's Guide to New Mathematics* (London, 1966 (original U.S. edition 1964)), pp. 1–2, 47–49, 71–72.

The Mathematics Revolution

If you have a child in school you may have begun to feel the impact of a change in mathematics—a change so far-reaching that it may well revolutionize the teaching of the subject at all levels and in many countries in the Western world, at least. ⌐...⌐ Some time in the middle 1950s it became abundantly clear to the U.S.A. that drastic modification of a limited and old-fashioned (non-progressive) syllabus in the schools was urgently needed. With determination and enthusiasm, research was undertaken and experiments were started. By 1965, great progress had been made; very large numbers of American schools had been able to introduce courses of study, even at an early age, aimed to develop a logical understanding of mathematics and the capacity to carry out thought processes of deductive reasoning. The aim was to avoid a purely automatic use of manipulative processes and learning by rote, with no comprehension of the principles involved. Naturally enough, word of the schemes crossed the Atlantic and, although the needs in Britain were not the same, ⌐...⌐ experiments began in various parts of the country, sometimes sponsored by university research workers and sometimes by schools. Needless to say, the teams soon linked up. It is, after all, difficult to introduce new approaches to teaching unless one has children able and willing to participate. At this stage there seemed to be two points of view which could be taken: on the one hand, it could be argued that there was already a comprehensive cross-section of mathematics in schools; on the other, that much of the existing curriculum of the average good school was unimaginative, stereotyped and archaic. One thing is, however, clear—that there was not such an urgent need for violent modification. The process has already started but it is, at present at any rate, a cautious and unhurried transformation.

❧

Properties of Number Systems: The ACD (associative, commutative, distributive) laws and arithmetic

The simplest of these laws, stated in mathematical language, is the commutative property for addition: $a + b = b + a$. For example, $5 + 1 = 1 + 5$. The commutative property for multiplication is $ab = ba$, eg $7 \times 3 = 3 \times 7$.

The law of association declares an equally obvious fact: that $2 + (3 + 4) = (2 + 3) + 4$. (Do the part in parentheses first and then add the other number.) Formally stated, the associative property for addition is: $a + (b + c) = (a + b) + c$. It is necessary because addition, like marriage, is a binary operation—only two can take part in it at a time. To add three numbers, you first add any two of them, then add that answer to the third one. The law of association merely states what you already know—that it does not matter which two you add first.

This property plays a vital role in the way children are now taught to add, as you can see in the following example:

$$9 + 3 = 9 + (1 + 2)$$

$$= (9 + 1) + 2$$

$$= 10 + 2$$

$$= 12.$$

They start with sets, regroup them by using the law of association, always building sets of ten, and then add as the union of sets. "We join sets, we add numbers," they say.

❧

A great deal hinges on the child's ability to split the second number into two parts, one of which will combine with the first number to make ten. There are pages of practice on this.

Often the way has been prepared by using, in kindergarten, the Cuisenaire rods. The system consists of a set of wooden pieces varying from one unit to ten units in length. Cuisenaire, a Belgian, used one centimetre as his unit. The system introduces the child to numbers through measuring, rather than by counting, the traditional gateway to arithmetic. If Billy wants to find $3 + 2$, instead of counting 3 clowns here and 2 clowns there, he places the 3-block end-to-end with the 2-block and then finds the block whose length measures the same.

A characteristic colour is used for each number; for example, the Cuisenaire rod for 10 is orange. The children learn to identify the block both by length and colour. If the blocks are piled up so that only part of the orange block is showing, Susie recognizes it as the 10-block by its colour, although she cannot see the full length.

The use of these rods may be carried on for some time in the child's school life. This accords with current pedagogical procedure, which is to present number concepts first through physical operations with three-dimensional objects, follow this with two-dimensional marks on paper, and then proceed to traditional Hindu–Arabic numerals.

Soon the children are handling sets of more than one ten—twenty, thirty, etc. (The objects in their sets have now lost some of their barnyard flavour—not so many baby ducks and frolicking lambs.) $22 + 4$ becomes $(20 + 2) + 4$, then $20 + (2 + 4)$ by the law of association, then $20 + 6$, and finally 26.

More complicated examples require the laws of both commutation and association.

$$23 + 30 = (20 + 3) + 30 \text{ (Because } 23 = 20 + 3)$$
$$= 20 + (3 + 30) \text{ (Addition is associative)}$$
$$= 20 + (30 + 3) \text{ (The commutative property permits}$$
$$\text{the 3 and the 30 to change places)}$$
$$= (20 + 30) + 3 \text{ (Addition is associative)}$$
$$= 50 + 3 \qquad (20 + 30 = 50)$$
$$= 53.$$

As progress is made the problems get harder:

$$38 + 3 = (30 + 8) + 3$$
$$= 30 + (8 + 3)$$
$$= 30 + (8 + [2 + 1])$$
$$= 30 + ([8 + 2] + 1)$$
$$= 30 + (10 + 1)$$
$$= (30 + 10) + 1$$
$$= 40 + 1$$
$$= 41.$$

True, nobody is going to go through life putting all this down whenever he wants to add. But this is a sound basic process on which to build. It is in fact easier, as skill develops, to add the tens, thus:

$$65 + 27 = (60 + 20) + (5 + 7)$$
$$= 80 + 12$$
$$= 92.$$

and later still

$$65 + 27 = 85 + 7$$
$$= 92.$$

Eventually an automatic mental process is evolved: 65, 85, 92.

"Merely a Formal Statement of the Way We Think"

Robert E. Eicholz and Phares G. O'Daffer, 1964

Dating, like the previous extract, from the beginning of the New Math era, Eicholz and O'Daffer's book, with set notation on page 1 and the whole numbers defined through set cardinality a few pages later, exemplifies the characteristics (and problems) of what was intended as a thoroughly systematic approach to learning mathematics. It aimed to give teachers a thorough understanding of the structure of mathematics underlying the New Math pedagogy. We have come a long way from Humfrey Baker's list of the parts of mathematics he proposed to teach, or indeed from the number games of Mrs. Lovechild's *Art of Teaching*.

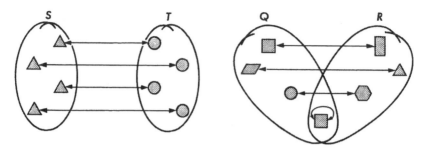

Figure 9.4. Sets S and T are disjoint; sets Q and R are not. (Eicholz et al., p. 6.)

Robert E. Eicholz and Phares G. O'Daffer, *A First Look at Modern Mathematics* (Palo Alto, 1964), pp. 6–7, 12–14.

Cardinal Numbers

If two sets S and T have the property that the elements of S can be paired with the elements of T so that each element in S is paired with exactly one element in T, and vice versa, then the sets S and T are said to be *equivalent* to each other. The relation shown by the pairing of elements is called a one-to-one correspondence between the sets.

Examples of pairs of equivalent sets are shown in Figure 9.4. Note that the sets S and T are disjoint with respect to each other, and that the sets Q and R are not disjoint. In the case of Q and R, we think of the square which is common to both sets as being paired with itself. Expanding upon this idea of pairing an object with itself, we point out that each set is equivalent to itself; that is, each element can be paired with itself.

With these observations as background, we can now make some statements about cardinal numbers. We shall not attempt a precise definition of the set of all cardinal numbers, but will confine ourselves to a simple description and some examples. Let us say that a cardinal number is a *class* of equivalent sets. For example, cardinal number 2 is the class of all sets which are equivalent to the set in Figure 9.5; cardinal number 3 is the class of all sets equivalent to the set $\{a, b, c\}$; etc. These statements are depicted in Figure 9.6. Cardinal number zero is the number of the empty set.

Since we think of each cardinal number as a *class* of sets, we can speak of *choosing* a set A *from* a cardinal number. If we choose a set from cardinal number 2, we have a set of two elements.

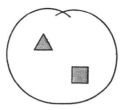

Figure 9.5. Cardinal number 2 is the class of all sets which are equivalent to this one. (Eicholz et al., p. 6.)

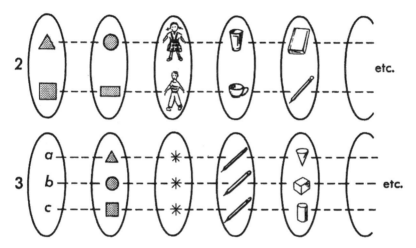

Figure 9.6. Cardinal numbers 2 and 3 as classes of equivalent sets. (Eicholz et al., p. 6.)

We have presented the idea of a class of equivalent sets in an abstract form. To explain this concept of number to a young child, we simply point repeatedly at representative sets and say the proper words. This is exactly how children are exposed to their first number concepts, and an understanding of the cardinal-number concept is well under way for most children by the time they enter school. For example, a child has heard the word "two" many times with respect to many different sets (two cows, two lollipops, two brothers) and has thus gained a feeling for the idea of "twoness."

Order for the set of cardinal numbers is easier to understand intuitively than it is to state formally. We state below precisely what it means to say that cardinal number *a* is greater than cardinal number *b*.

Consider two cardinal numbers, a and b, and sets from these cardinal numbers, A and B respectively. If the set A contains a proper subset which is equivalent to B, then we say that cardinal number a is greater than cardinal number b, and that cardinal number b is less than cardinal number a. We write $a > b$ (a is greater than b) and $b < a$ (b is less than a).

Careful examination of this definition will reveal that it is merely a formal statement of the way we think when we compare cardinal numbers.

❧

Addition of Cardinal Numbers

We are now in a position to say exactly what is meant by the sum of two cardinal numbers.

Consider two cardinal numbers a and b and sets A and B from these cardinal numbers such that $A \cap B$ is empty. (A and B are disjoint.) The cardinal number of the set $A \cup B$ is the *sum* of cardinal numbers a and b (written $a + b$). The numbers a and b are called *addends* of the sum $a + b$.

An example illustrates that this definition is little more than a formal statement of how we introduce children to the concept of addition. Suppose we wish to show a child that $2 + 3 = 5$. We would probably begin by displaying a set of 2 and a set of 3 and having the child identify the numbers of these sets. (Select a set from cardinal 2 and a set from cardinal 3.) Now we would have the child push the two sets together. (Form the union.) We would ask, "How many in all?" (The sum is the cardinal number of the union.)

Observe the importance of the word *disjoint* in the definition of the sum of two cardinal numbers. Suppose we wish to demonstrate the sum of 4 and 5 and fail to select disjoint sets.

$$\text{Set of 4} \longrightarrow \{a, b, c, d\}.$$
$$\text{Set of 5} \longrightarrow \{c, d, e, f, g\}.$$
$$\text{Union of the sets} \longrightarrow \{a, b, c, d, e, f, g\}.$$

The cardinal number of the union is 7, and this is not the number that we wish to call the sum of 4 and 5. Obviously, in order to obtain the desired result, we must select disjoint sets.

Some special remarks will be helpful.

(1) We add numbers, not sets.

(2) We form the union of sets, not of numbers.

(3) If A and B are sets, $A = B$ means that A and B are the same set; that is A and B are different names for the same set. If A and B are equivalent but different sets, we do *not* write $A = B$.

(4) If a and b are cardinal numbers, and we write $a = b$, we mean that a and b are the same cardinal number; that is, a and b are different names for the same cardinal number.

(5) *Greater than* and *less than* are relations defined for numbers, not for sets. We do not write $A > B$ or $B < A$ for sets A and B.

Exercises

1. From the definition of union of sets, we can see that $A \cup B$ is the same set as $B \cup A$. What does this fact tell you about the cardinal numbers $a + b$ and $b + a$?

2. Suppose that a, b, and c are cardinal numbers such that $a + b = c$. What can you conclude if $a = c$? if $c > a$?

3. Show a set from cardinal 4 and a set from cardinal 5 such that the cardinal number of the union of these two sets is 5. Make choices so that the cardinal number of the union is 6; so that it is 9. Can sets be chosen from 4 and 5 so the cardinal number of the union is greater than 9? less than 5?

4. If the cardinal number of the union of two sets is less than the sum of the cardinal numbers of the sets, what can you say about intersection of the two sets?

5. Consider three cardinal numbers a, b, and c such that $a + b = c$ and $a + c = b$. What can you say about cardinal number a?

6. At a school picnic, 16 children drank cherry soda and 15 had lemon soda. There were only 25 children at the picnic. Explain in set language why one cannot argue that there were 31 children at the picnic.

7. List the six possible pairs of addends of 7.

Turtle Fun

Serafim Gascoigne, 1985

Many readers will remember the LOGO turtle, but it comes as something of a surprise to recall that it was described in the 1980s as representing an entirely new form of geometry. As we saw in Chapter 6, the late twentieth century saw the rise of computers and calculators as the new "mathematical instruments," and children received instruction in their use just as adults did.

Nothing illustrates this better than these passages about the turtle and its antics. The same book also has a wonderful passage about another piece of microchip wildlife, the "mouse": "The busy executive can sit back in his leather swivel chair and call all sorts of information on to the screen simply by moving and squeezing the mouse!"

You can try out the instructions given here using one of the many turtle programs on the web, such as www.mathsnet.net/logo/turtlelogo.

Serafim Gascoigne (dates unknown), *Turtle Fun: LOGO for the Spectrum 48K* (London, 1985), pp. vii, 6–8, 34–41, reproduced with permission of Palgrave Macmillan.

The world of microcomputers is forever changing. There is always something new to learn and find out. How can you keep up? Will you be able to program the computers of the future? And what about "intelligent" machines and robots, especially the LOGO turtle?

This book will help you to keep up with events by introducing you to a very powerful control language of the future, called LOGO. LOGO has in fact been developed by computer scientists working in a new science called Artificial Intelligence. LOGO has been written for young programmers. It comes from another language called LISP, which is one of the main languages used to control "intelligent" computers and robots. Using such commands as STARTROBOT or REPEAT [FORWARD 50 LEFT 120] for example, you can either drive a small mechanical device called the floor turtle (see Figure 9.7) or control the screen turtle, a movable cursor on your TV screen. You can also teach the turtle your own commands, called "procedures," which can include graphics, text and sound.

Figure 9.7. A small mechanical device called the floor turtle. (Gascoigne, p. 6.) Reproduced with permission of Palgrave Macmillan.

Learning LOGO will not only teach you another computer language but it will also help you to organise your thinking and ideas in a clear and structured way. Apart from the scientific value of learning LOGO, it is fun! The world of the LOGO turtle is an exciting and intriguing world that ranges from simple screen doodling to the study of turtle geometry—a new geometry, totally different from school geometry but surprisingly familiar!

LOGO allows you to invent your own language, to be creative and above all, to make your programming a personal venture.

How to move the turtle

Before you begin to explore the ideas in this book, you will need to know how to use some basic commands. If you already know how to move the turtle, you may skip this section.

⌐...⌐ To call up the turtle, you type SHOWTURTLE (ST). You will now see a turtle or cursor-shape at the centre of the screen. This is the LOGO screen turtle.

To move the turtle you use the following commands

FORWARD   RIGHT
BACK   LEFT

These commands can be shortened to FD (FORWARD), BK (BACK), RT (RIGHT) and LT (LEFT). Most LOGO commands that you will meet in this book can in fact be shortened.

Typing FORWARD 10 or FD 10 (and pressing ENTER) will move the turtle forward ten steps on the screen, and FORWARD 20 or FD 20 (press ENTER) will move the turtle forward twenty steps.

Typing BACK 10 or BK 10 (press ENTER) will move the turtle backwards ten steps, and so on.

Try changing the number of steps after the commands, FORWARD and BACK.

To turn the turtle to the left you type: LEFT followed by a number. This is the number of degrees. For example LEFT 90 or LT 90 (press ENTER) will turn the turtle to the left 90 degrees. (If you are not sure what a degree is, it is a unit of measurement like miles or litres. Degrees are used to measure angles just as miles are used to measure distance or litres are used to measure quantities of liquid.)

RIGHT 90 or RT 90 (press ENTER) will turn the turtle to the right 90 degrees. LEFT 1 or LT 1 will turn the turtle to the left 1 degree.

RIGHT 45 or RT 45 will turn the turtle to the right 45 degrees.

Try changing the number of degrees after the commands, LEFT and RIGHT. Don't forget to press ENTER whenever you want the computer to carry out your commands.

As the turtle moves, you will have noticed that it draws a line. If you wish, you can move the turtle without drawing a line by typing PENUP or PU. When you wish to resume drawing, type PENDOWN or PD.

You can also hide the turtle as it draws by typing HIDETURTLE. To make the turtle reappear simply type SHOWTURTLE.

The basic commands so far are

FORWARD	BACK	RIGHT
PENUP	PENDOWN	LEFT
HIDETURTLE	SHOWTURTLE	

One of the important things to do with LOGO is to experiment! Try out these commands and see what happens. Some important ideas have been discovered by people doodling on the screen. From the very moment that you start moving the turtle, you are ready to discover new ideas. See what you can find.

Pretty poly

You may have enjoyed drawing polygons at school using a ruler and pencil. Even if you did not enjoy drawing them, using the turtle to draw polygons can be an exciting experience. The power of the turtle allows you to produce some spectacular effects at the press of a button. Figures that would require hours of precision drawing with a pencil can be produced on the screen instantly.

Here is a simple procedure to get you started. You can draw polygons of all shapes and sizes. The procedure, which I have called POLY, is as follows

```
TO POLY :SIZE :ANGLE
FD :SIZE
LT :ANGLE
POLY :SIZE :ANGLE
END
```

By typing POLY, followed by an input for the sides and an input for the angle, you can produce some amazing figures on the scceen. The last line in the procedure is recursive. It tells the turtle to start again.

You might like to experiment with the ANGLE input. Can you make any predictions as to what kind of shape the turtle will produce? How many sides will it have? How many points or vertices? Will the lines cross each other?

POLY with subprocedures

You can also use subprocedures inside the POLY procedure. Instead of FD, for example, use the name of a procedure such as TRIANGLE or SQUARE.

To demonstrate this, let's write a new version of POLY and call it POLY1.

```
TO POLY1 :SIZE :ANGLE
TRIANGLE :SIZE
RT :ANGLE
POLY1 :SIZE :ANGLE
END
```

If you have not already got a procedure called TRIANGLE, here is one for you to experiment with.

```
TO TRIANGLE :SIZE
REPEAT 3 [FD :SIZE RT 120]
END
```

You can of course invent your own program to draw a triangle. Try

POLY1 60 144

Polyspirals

From the simple POLY procedure you can go on to produce what are called polyspirals. For this let's create a new procedure and call it POLYSPI.

```
TO POLYSPI :SIZE :ANGLE
FD :SIZE
LT :ANGLE
POLYSPI :SIZE + 3 :ANGLE
END
```

This new procedure is really the POLY procedure again. However the last line

POLYSPI :SIZE + 3 :ANGLE

tells the turtle to increase the size of the figure by 3 each time. The increase in size is written in the last line, which also uses recursion to repeat POLYSPI.

You can go a stage further and make the increase in size (called an increment) a variable as well. Let's call it INC.

```
TO POLYSPI1 :SIZE :ANGLE :INC
FD :SIZE
LT :ANGLE
POLYSPI1 :SIZE + :INC :ANGLE :INC
END
```

This procedure allows you to chooose the increment or increase in step each time. However the increment remains the same throughout the procedure. It is the :SIZE that gets bigger. Note that in the last line of this procedure :INC has to be written twice! POLYSPI1 needs three inputs. The combination :SIZE + :INC is regarded as one input for :SIZE. Therefore :INC is written again in the same line.

The next procedure, called POLYSPI2, makes the increment grow this time in proportion to :SIZE.

```
TO POLYSPI2 :SIZE :ANGLE
FD :SIZE
LT :ANGLE
POLYSPI2 :SIZE + :SIZE/10 :ANGLE
END
```

Here you can see that the increment is not written as a variable but is included in the variable :SIZE. In the recursion line I have written

:SIZE + :SIZE/10

This simply means, add the value of :SIZE to :SIZE divided by 10.

You can carry out any arithmetical operations you like on input. For example, you can use +, −, * and /. (* is used for multiply and / is used for divide.) Have a go at using different operators inside your procedures. Here are just a few examples.

```
FD :SIZE * 4     (multiplication)
FD :SIZE / 2     (division)
FD :SIZE + 10    (addition)
```

You can even use operators with the inputs to your procedures. You can type, for example

POLYSPI2 3 360/7

where you are asking the turtle to divide the angle of 360 degrees by 7.

Here is another program that uses BK as well as FD and uses + and / as operators.

```
TO POLYSPI3 :SIZE :ANGLE :INC
WINDOW SETBG 1 SETPC 3
FD :SIZE SETPC 7 FD 20 + :SIZE/2
BK :SIZE/2 RT :ANGLE
POLYSPI3 :SIZE + :INC :ANGLE :INC
END
```

POLYSPI3   20   162   5

            20   160   1

            20   119   3

             3   110   3

             1   72   1

             3   75   1

             3   360/7   2

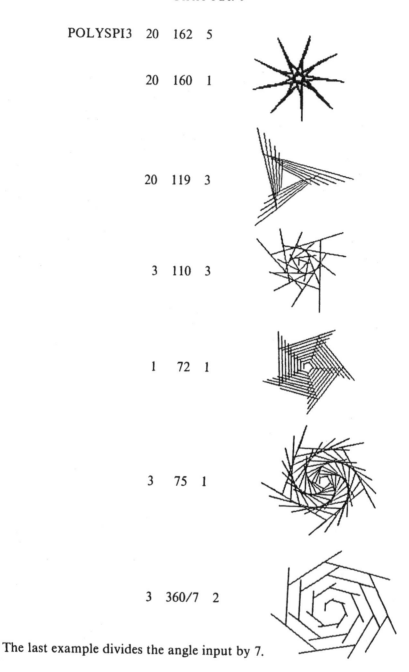

The last example divides the angle input by 7.

Figure 9.8. Turtle frenzy: spirals and polyspirals. (Gascoigne, p. 38.) Reproduced with permission of Palgrave Macmillan.

I have used some cosmetics here! You can set your own background and pen colour. As you can see, the size inputs are divided by 2, except in the last line.

Here (see Figure 9.8) is a list of inputs that you might like to try.

Stopping your procedures

You will have noticed that POLY and POLYSPI do not stop. This is because we have used recursion. To stop a procedure that is recursive you need a conditional statement.

IF :SIZE > 100 [STOP]

This tells the turtle to stop if the size of the figure is greater than 100. This conditional statement using IF can be used with any procedure in which you are using recursion.

Using RANDOM

Here is another poly program that uses RANDOM.

```
TO RANDP :SIZE :ANGLE :INC
MAKE "ANS RANDOM 10
IF :ANS = 0 [PD][PU]
FD :SIZE LT :ANGLE
RANDP :SIZE + :INC :ANGLE :INC
END
```

The conditional statement in this procedure tells the turtle to pendown if the random number is 0. If not, then penup.

❧ 10 ❧

"So Fundamentally Useful a Science": Reflections on Mathematics and Its Place in the World

GALILEO FAMOUSLY CLAIMED THAT THE BOOK OF NATURE IS WRITTEN IN THE language of mathematics, and both before and since his time mathematicians and philosophers have tried to work out the details, the limits, and the meaning of the applicability of mathematics. We have seen, in Chapter 8, mathematics at work in various contexts; in this chapter we see mathematics being used more publicly: in politics or large-scale astronomy, in early modern natural philosophy, and in modern physics. In parallel we see its ever more public role being reflected upon.

One strand of those reflections is a particular flavor of optimism about the achievements of a mathematical world view. Many writers have concurred with the quote from Joseph Glanvill (1664) in the title of this chapter, saying that mathematics is "fundamentally useful"; 350 years after Galileo we see Richard Feynman echoing the same claim and expounding the idea that the world *can only* be understood through mathematics.

Another strand is the tendency to think of logic—and even reasoning in general—as parts of mathematics: Cassius Jackson Keyser, in 1929, could give a simple piece of deduction as an example of mathematics.

There have been critics, too, of mathematical ways of doing things, and we will see Sylvester fighting off Huxley in 1870, and Allen Hammond arguing in the 1970s for a more humane view of mathematicians and their work.

Whether or not we agree with the highest flights of optimism or the boldest claims for what mathematics is, there is little denying its pervasive effectiveness in the modern world—or the depth of the reflections to which it has given rise.

The Myrrour of the Worlde

Gossuin of Metz, 1481

The remarkable book entitled *Image du monde* appeared in about 1246; it is normally attributed to the French priest and poet Gossuin of Metz. Written in French verse, the book included both an account of the Biblical creation and discussions of the shape and size of the cosmos, as well as descriptions of some of its contents, forming a sort of compendium of what one ought to know about the world. It received translations into several languages, and it caught the eye of William Caxton, the first English printer. He made his English translation of the book in 1480 and printed it the following year as *the myrrour of the worlde*.

The section given here describes the final four of the "seven liberal arts"— arithmetic, geometry, music, and astronomy (the first three are grammar, logic, and rhetoric)—and explains why they are important. It is one of the very first discussions of mathematical subjects—albeit in rather general terms—to be printed in English. Caxton's prose is lightly paraphrased here.

Gossuin of Metz (thirteenth century), trans. William Caxton (c. 1422–1491), *The myrrour of the worlde* ... (Westminster, 1481), c5^r–c7^v.

Here followeth Arithmetic, and whereof it proceedeth

The fourth science is called arithmetic; this science cometh after rhetoric, and is set in the middle of the seven sciences. And without her may none of the seven sciences perfectly—nor well and entirely—be known; wherefore it is expedient that it be well known and conned. For all the sciences take of it their substance, in such a way that without her they may not be. And for this reason was she set in the middle of the seven sciences, and there holdeth her number. For from her proceed all manner of numbers, and in all things run, come and go. And nothing is without number. But few perceive how this may be, unless he have been master of the seven arts so long that he can truly say the truth. But we may not now recount nor declare all the causes wherefore; for he that would dispute upon such works, to him it behooves to dispute and to know many things ⌐...⌐. Who that knew well the science of arithmetic, he might see the ordinance of all things. By ordinance was the world made and created, and by ordinance of the sovereign it shall be defeated.

Next followeth the science of Geometry

The fifth is called geometry, the which more availeth to Astronomy than any of the seven others; for by her is compassed and measured Astronomy. Thus are, by geometry, measured all things where there is measure. By geometry may be known the course of the stars, which always go and move, and the greatness of the firmament, of the sun, of the moon and of the earth. By geometry may be known all things, and also the quantity. They may not be so far—if they may be seen or espied with eye—but it may be known. Who well understood geometry, he might measure in all mysteries; for by measure was the world made, and all things high, low and deep.

Here followeth of music

The sixth of the seven sciences is called music, which formeth himself of Arithmetic. Of this science of music cometh all temperament; and of this art proceedeth some physic. For like as music accordeth all things that discord in themselves, and returns them to concordance; right so in like ways worketh physic to bring Nature to point that disnatureth in man's body, when any malady or sickness encumbereth it.

~❖~

By her the seven sciences were set in concord, that they yet endure. By this science of music are extracted and drawn all the songs that are sung in holy church, and all the accordances of all the instruments that have diverse accords and diverse sounds. And where there is reason and understanding of some things, truly, he who knows well the science of music, knoweth the accordance of all things. And all the creatures that take pains to do well remain in concordance.

Here speaketh of Astronomy

The seventh and the last of the seven sciences liberal is astronomy, which is of all clergy the aim. By this science may, and ought to be, enquired of things of heaven and of the earth, and especially of them that are made by nature, how far that they be. And who knoweth well and understandeth astronomy, he can set reason in all things. For our creator made all things by reason, and gave his name to every thing.

By this Art and science were first emprised and gotten all other sciences of decrees and of divinity, by which all Christianity is converted to the right faith of our lord God to love him and to serve the king almighty, from whom all goods come and to whom they return, which made all astronomy, and heaven and earth, the sun, the moon and the stars, as he that is the very ruler and governor of all the world, and he that is the very refuge of all creatures, for without his pleasure nothing may endure. Truly, he is the very Astronomer, for he knoweth all the good and the bad as he himself that composed astronomy, that sometime was so strongly frequented, and was held for a right high work. For it is a science of such noble being that who that might have the perfect knowledge thereof, he might well know how the world was compassed, and plenty of other partial sciences. For it is the science above all others, by which all manner of things are known the better.

By the science of Astronomy only, were founded all the other six aforenamed. And without them may none know aright Astronomy, be he never so sage or mighty. In like wise as a hammer or another tool of a mason be the instruments by which he formeth his work, and by which he doth his craft, in like wise by right mastery be the others the instruments and fundaments of Astronomy.

"A Very Fruitfull Praeface"

John Dee, 1570

One of the classic descriptions of what mathematics is and does is the preface written by John Dee for Billingsley's translation of Euclid; for an extract from the translation itself, see Chapter 7. It contains a remarkable mix of factual information, classical quotation, metaphysical speculation, and what seems to be straightforward invention about the parts of mathematics and their relationships. Dee himself was a remarkable polymath, notorious in his own day for his (alleged) conversations with angels.

The text presented here is slightly paraphrased.

John Dee (1527–1609), "Praeface," in *The elements of geometrie of the most aunancient philosopher Euclide of Megara. Faithfully (now first) translated into the Englishe toung, by H. Billingsley, citizen of London. Whereunto are annexed certaine scholies, annotations, and*

inuentions, of the best mathematiciens, both of time past, and in this our age. With a very
fruitfull praeface made by M. I. Dee, specifying the chiefe mathematicall scie_nces, what they
are, and wherunto commodious: where, also, are disclosed certaine new secrets mathematicall
*and mechanicall, vntill these our daies, greatly missed. (London, 1570), iv^v–*i^r.*

All things which are, and have being, are found under a triple general division.

For either they are deemed Supernatural, Natural, or of a third being. Things Supernatural are immaterial, simple, indivisible, incorruptible, and unchangeable. Things Natural are material, compounded, divisible, corruptible, and changeable. Things Supernatural are comprehended by the mind only; things Natural are able to be perceived by the exterior senses. In things Natural, probability and conjecture have place; but in things Supernatural, demonstration and most sure knowledge is to be had. By which properties and comparisons of these two, more easily may be described the state, condition, nature and property of those things which we before termed of a third being, which, by a peculiar name, also are called Things Mathematical.

For these, being (in a manner) middle, between things supernatural and natural, are not so absolute and excellent as things supernatural, nor yet so base and gross as things natural. But they are things immaterial, and, nevertheless, by material things are able, somewhat, to be signified. And though their particular Images, by Art, may be aggregated and divided, yet the general Forms, notwithstanding, are constant, unchangeable, untransformable, and incorruptible. Neither can they, at any time, be perceived or judged by the senses, nor yet, for all that, conceived in the royal mind of man. But, surmounting the imperfection of conjecture, thinking and opinion, and coming short of high intellectual conception, they are the Mercurial fruit of intellectual discourse, subsisting in perfect imagination.

A marvellous neutrality have these things Mathematical, and also a strange participation between things supernatural, immortal, intellectual, simple and indivisible, and things natural, mortal, sensible, compounded and divisible. Probability and sensible proof may well serve in things natural, and are commendable; in Mathematical reasonings, a probable Argument is nothing regarded, nor yet the testimony of sense any whit credited. But only a perfect demonstration, of truths certain, necessary, and

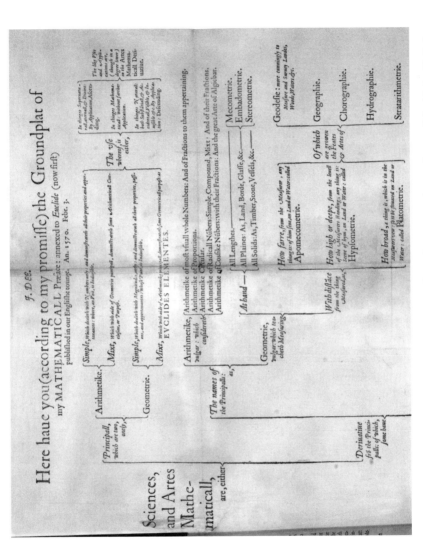

Figure 10.1. Part of Dee's "Groundplat," showing the parts of mathematics. (Dee, facing fol. Aiv^v. © Princeton University Library. Rare Books Division, Department of Rare Books and Special Collections, 2654.331.570q.)

invincible, universally and necessarily concluded, is allowed as sufficient for an Argument exactly and purely Mathematical.

Of Mathematical things there are two principal kinds: namely Number and Magnitude. Number we define to be a certain Mathematical Sum of Units. And an Unit is that thing Mathematical and Indivisible, by participation in some likeness of whose property any thing which is indeed—or is counted—One, may reasonably be called One. We account an Unit a thing Mathematical, though it be no Number, and also indivisible, because, materially, Number doth consist of it, which, principally, is a thing Mathematical.

Magnitude is a thing Mathematical, by participation in some likeness of whose nature any thing is judged long, broad, or thick. A thick Magnitude we call a Solid, or a Body. What Magnitude soever is Solid or Thick is also broad and long. A broad magnitude we call a Surface or a Plane. Every plane magnitude hath also length. A long magnitude we term a Line. A Line is neither thick nor broad, but only long.

Every certain Line hath two ends; the ends of a line are called Points. A Point is a thing Mathematical and indivisible, which may have a certain determined position. If a Point move from a determined position, the path wherein it moves is also a Line, mathematically produced. Whence, by the ancient Mathematicians, a Line is called the course of a Point. A Point we define by the name of a thing Mathematical, though it be no Magnitude, and indivisible, because it is the proper end and bound of a Line, which is a true Magnitude. And Magnitude we may define to be that thing Mathematical which is divisible for ever, in parts divisible, long, broad or thick. Therefore, though a Point be no Magnitude, yet we reckon it a thing Mathematical (as I said), since it is properly the end and bound of a line.

"Geometry Is Improving Daily"

Joseph Glanvill, 1664

In his spirited defense of modern scientific learning in general and the young Royal Society in particular (contrast this with Margaret Cavendish's satire in Chapter 11), the Church of England clergyman Joseph Glanvill somewhat unexpectedly included a section on mathematics. The subject was not central to the Royal Society's concerns at the time, and Glanvill's discussion is fascinating both for his belief in the importance of mathematics (a "mighty help," "fundamentally useful") and for his awareness of the processes of historical change within the discipline. Both of these characteristics are on display in these excerpts.

Joseph Glanvill (1636–1680), *Plus ultra, or, The Progress and Advancement of Knowledge since the Days of Aristotle* (London, 1668), pp. 19–38.

I Proceed now to my third Instance of arts, ⌐...⌐ which are Advantages for deep search into Nature, and have been considerably advanced by the Industry and culture of late Times, above their ancient Stature. And the Instance was,

The Mathematics

That these are mighty helps to practical and useful Knowledge, will be easily confessed by all that have not so much ignorance as to render them incapable of information in these matters. And the Learned Gerard Vossius° hath proved it by induction in particulars. And yet it must be acknowledged that Aristotle, and the disputing Philosophers of his School, were not much addicted to those noble Inquisitions. For Proclus, the Commentator upon Euclid, though he gives a very particular Catalogue of the Elder Mathematicians, yet hath not mentioned Aristotle in that number. And though Diogenes Laertius° takes notice of a Book he inscribed ⌐*Mathematikon* (Mathematics)⌐, another, ⌐*Perì monádōs* (On Unity)⌐, and a Third, yet extant, ⌐*Perì 'atópōn grammōn* (On strange diagrams)⌐, yet it appears not that these were things of very great value. And Aristotle's Metaphysical procedure, even in Physical Theories, the genius and humour of his Principles, and the airy contentions of his Sect, are huge presumptions that this Philosopher was not very Mathematical. And his numerous succeeding Followers were certainly very little conversant in those generous Studies. I have elsewhere taken notice, that there is more published by those Disputing men on some paltry trifling Question about *ens Rationis* ⌐(mental being)⌐, and their *Materia prima* ⌐(first material)⌐, than hath been written by their whole number upon all the vast and useful parts of Mathematics and Mechanics. There was a time when these were counted Conjurations; and I do not very well know the reason of ⌐my

opponent's͵ displeasure at my Discourse about Dioptric Tubes, ͵...͵ except
he was under the dread of some such fancy, and believed there was Magic
in Optics. It would require much skill in those Sciences to draw up the
full History of their Advancements; I hear a very accurate Mathematician
is upon it, and yet to fill up my Method, I'll adventure at some imperfect
Suggestions about the Inventions and Improvements of this kind. And I
begin with

Arithmetic,

which is the handmaid to all the other parts of Mathematics. This indeed
Pythagoras is said to have brought from the Phoenicians to the Grecians,
but we hear no great matter of it till the days of Euclid: not the Euclid that
was the Contemporary of Plato, and Hearer of Socrates, but the famed
Mathematician of that Name, who was after Aristotle, and at 90 years
distance from the former. This is the first Person among the Ancients that
is recorded by the exact Vossius to have done anything accurately in that
Science.

After him it was advanced by Diophantus, methodized by Psellus,
illustrated among the Latins by L. Apuleius, and in later times much
promoted by Cardan, Gemma Frisius, Ramus, Clavius,° and diverse more
modern Artists, among whom I more especially take notice of that Ingenious Scot the Lord Napier, who invented the Logarithms, which is
a way of computing by artificial Numbers, and avoiding the tedium of
Multiplication and Division. For by this Method all those Operations are
performed by Addition and Subtraction, which in natural Numbers were to
be done those longer ways. This Invention is of great use in Astronomical
Calculations, and it may be applied also to other Accounts. Besides this,
the same Learned Lord found an easy, certain, and compendious way of
Accounting by Sticks, called Rabdology, as also Computation by Napier's
Bones. Both these have been brought to greater perfection by others, since
their first Discovery, particularly by Ursinus and Kepler.

To them I add the Decimal Arithmetic, which avoids the tedious way of
computing by Vulgar Fractions in ordinary Accounts, and Sexagenaries
in Astronomy, exceedingly and lately improved by our famous Oughtred,
and Dr. Wallis° a Member of the Royal Society. If I should here subjoin
the Helps this Art hath had from the Works and Endeavours of Anatolius,

Barlaam, Maximus Palanudes, Nemorarius, Florentinus Bredonus, Pisanus, Orentius, and in this Age, from those of Adrianus Romanus, Henischius, Cataldus, Malapartius, Keplerus, Briggius, Crugerus, and a vast number reckoned up by Vossius,° I should be tedious on this Head; and therefore I pass lightly over it, and proceed to

Algebra,

of universal use in all the Mathematical Sciences, in Common Accounts, in Astronomy, in taking Distances and Altitudes, in measuring plane and solid Bodies, and other useful Operations. The first noted Author in this Method was Diophantus, who lived long since the Idol of Disputers.° He, and those other Ancients that used it, performed their Algebraical Operations by Signs and Characters suited to the several Numbers, and powers of Numbers, which they had occasion to use in solving Problems. But the later Mathematicians have found a far more neat and easy way, *viz.* by the Letters of the Alphabet, by which we can solve many Problems that were too hard for the Ancients, as far as can be discovered by any of their remaining Works. For there were many affected Equations (as they call them) that did not equally ascend in the Scale of Powers, that could not be solved by the elder Methods; whereas the acute Vieta,° a Mathematician of this last Age, affirms, he could resolve any Problem by his own Improvements. Besides him, our excellent Oughtred, another, lately mentioned, did much in this way. But the inimitable Descartes hath vastly out-done both former and later Times, and carried Algebra to that height, that some considering men think Human Wit cannot advance it further. I will not say so much, but no doubt he hath performed in it things deserving vast acknowledgment, of which you shall hear more anon. And from hence I step to the Consideration of

Geometry,

which is so fundamentally useful a Science that without it we cannot in any good degree understand the Artifice of the Omniscient Architect in the composure of the great World, and our selves. ⌞*Theòs geometrēi* (God geometrizes)⌟ was the excellent saying of Plato; and the Universe must be known by the Art whereby it was made.

Notes

Gerard Vossius (1577–1649): humanist, classical scholar, and theologian who wrote A *Book on the Nature and Constitution of the whole of mathematics*, published posthumously in 1650; it was one of the century's most important contributions to the history of mathematics.

Diogenes Laertius: third-century Greek writer on philosophy.

Psellus and *Apuleius* wrote, respectively, in the eleventh and the second centuries; *Cardan, Gemma Frisius, Ramus*, and *Clavius* were sixteenth-century scholars and mathematicians.

Oughtred: William Oughtred (1574–1660), an important writer on algebra also remembered for inventing a kind of slide rule. John Wallis (1616–1703) was the Savilian Professor of Geometry at Oxford for more than fifty years, and one of the most important British mathematicians of his day.

Anatolius …: Not to be tedious, these were all writers on arithmetic or, later, algebra, ranging from Greek authors of the third century (Anatolius) through medieval Latins (Nemorarius) to seventeenth-century Britons (Briggius).

Idol of Disputers: Aristotle.

Vieta: Françis Viète (1540–1603), French mathematician and astronomer; he made very important innovations in algebraic notation and famously claimed that his methods would "leave no problem unsolved."

The Fifth Element
Edmund Scarburgh, 1705

We have already met Edmund Scarburgh and his Euclid in Chapter 7; here we see him introducing Book 5 of the *Elements*. On the face of it, that book is about rather abstract matters of ratio and the manipulation of ratios; but, in a passage which reflects the self-confidence of British mathematical natural science in the wake of Newton's *Principia*, Scarburgh interprets the book as "an universal Mathesis," a vastly general set of methods of use across the whole range of scientific studies. He also expounds briefly on what the student needs in order to understand these treasures, and thus by implication on what a proper approach to natural science itself must consist of: a thorough grounding in mathematics, starting with arithmetic.

Edmund Scarburgh, *The English Euclide, being The First Six Elements of Geometry, Translated out of the Greek, with Annotations and useful Supplements* (Oxford, 1705), p. 176.

⌊Introduction to⌋ The Fifth Element

This element depends upon none of the foregoing, but stands alone as an *universal Mathesis*.° It is like Metaphysics to Natural Philosophy, a Transcendent Element of pure and prime Mathematics, and so much abstracted not only from Matter in any Subject, but also from every particular kind of Subjects, so as to be equally applicable to all the Species of Quantity, to the Sciences, Geometry, and Arithmetic, and besides, universally to all other things which are capable of comparison, such as Force and Power in Agents, Intension and Remission in Qualities, Velocity and tardity in Motions, gravity and Levity in Ponderations, Modulation in sounds, Value and Estimation in Things, and whatsoever else may admit of any Gradation.

But Euclid in a geometrical method pursues his course, and does accordingly apply this Element to Magnitudes, yet in such an artificial and subtle Form of Demonstration that it might in general be made use of wheresoever in the nature of things the reason of Man can compare one thing with another.

This Doctrine of Proportions cannot be well explained without the use of Numbers; and therefore whoever intends rightly to understand this Element must come furnished with a moderate skill in Arithmetic. We have therefore applied Numbers to the Definitions and Propositions, for illustration's sake to the younger students.

I should farther advise that with the Study of this Element, also Euclid's Elements of Numbers were together perused, especially those Propositions where Proportions are concerned. For the Doctrine of Proportions is chiefly, or rather only, explicable by Numbers, and what here is applied to Magnitudes was secretly derived from those Elements, which do much further a right understanding of this. It will be at first sufficient for Beginners only to read the Propositions of those Elements, and carefully to observe the Expositions, which may instruct them enough for their present use in this Element, without giving themselves the trouble of being convinced by Demonstrations.

Note

Mathesis: a (mathematical) science.

Of Mathematics in General

Richard Sault, 1710

We met the *Athenian Mercury* and its mathematical author Richard Sault in Chapter 5; when its contents were reprinted in book form early in the eighteenth century, they were supplemented with some general discussion of the subjects treated, including mathematics. This passage, with its view of mathematics as chiefly a practical help to making illusions and entertainments, makes an interesting comparison with the view of mathematics and its history displayed by Joseph Glanvill fifty years earlier.

Richard Sault (d. 1702), "Of mathematics in general" in *A Supplement to the Athenian Oracle: Being a Collection of the Remaining Questions and Answers in the Old Athenian Mercuries. Intermixt with many Cases in Divinity, History, Philosphy, Mathematics, Love, Poetry, never before Publish'd. To which is prefix'd The History of the Athenian Society, And an Essay upon Learning. By a Member of the Athenian Society.* (London, 1710), pp. 95–96.

To speak a little of mathematics in general, before we come to treat of any particular Parts of that Subject, we suppose we cannot do better than to give a short account of what has been already performed by the assistance of this Art, that we may the better judge of the Possibility of future Acquirements. We read of many Persons, who in this Study have trod so near upon the heels of Nature, and dived into things so far above the Apprehension of the Vulgar, that they have been believed to be Necromancers, Magicians, etc., and what they have done to be unlawful, and performed by Conjuration and Witchcraft, although the fault lay in the People's Ignorance, not in their Studies. But to the Instances we promised.

Regiomontanus's Wooden Eagle and Iron Fly, mentioned by Petrus Ramus, Hakew, Heylin,° etc., must be admirably contrived, that there was so much Proportion, such Wheels, Springs, etc., as could so exactly imitate Nature. The first was said to fly out of the City of Noremberg, and meet the Emperor Maximilan, and then returned again, waiting on him to the City Gates. The other, to wit, the Fly, would fly from the Artist's Hand round the Room, and return to him again. This Instance proves the Feasibility of doing things of great use, as that Action of Proclus the mathematician,° in the Reign of Anastasius Dicorus, who made Burning-Glasses with that skill and admirable force, that he therewith burnt, at a great distance, the Ships of the Mysians and Thracians, that blocked up the City of Constantinople.

We shall pass over the Curiosities and admirable Inventions, which are mentioned in the Duke of Florence's Garden at Pratoline, as also those of the Gardens of Hippolitus d'Este, Cardinal of Ferrara, at Tivoli near Rome, because they were more design'd for Pleasure, than real Use. For our Design is only to show the real Advantage that may be drawn from Mathematics; though we are also certain, that the most surprising Pleasures of Nature depend upon it.

The great clock of Copernicus° was certainly a curious Masterpiece, which showed the Circuitions of all the Celestial Orbs, the Distinction of Days, Months, and Years, where the Zodiac did explicate its Signs, the Changes of the Moon, her Conjunctions with the Sun; every Hour produced upon the Scene some Mystery of our Faith, as the first Creation of Light, the powerful Separation of the Elements, etc. What shall we say of Cornelius Van Drebble's Organ,° that would make an excellent Symphony itself, if set in the Sunshine in the open Air? Or of Galileo's imitating the Work of the first Day: *fiat lux*, Let there be Light? Or of Granibergius's Statue, which was made to speak? Or in fine, of that Engine at Danzig in Poland, which would weave four or five Webs, all at a time, without any human Help? It worked night and day, but was suppressed, because it would have ruined the poor People. These few Instances give a rude Prospect of what one may probably expect from a due Application of the Mind to the Study of Mathematics.

Notes

Regiomontanus's Wooden Eagle and Iron Fly: Johannes Müller von Königsberg (1436–1476), known as Regiomontanus, was a German mathematician and astronomer. The stories of his marvelous automata told by Petrus Ramus (1515–1572) seem, unfortunately, rather doubtful.

Proclus the mathematician: Proclus Diadochus (411–485) is remembered as a writer of mathematical commentaries.

The great clock of Copernicus: Sault probably refers to the famous clock in Strasbourg Cathedral, completed in 1574, which showed, among other things, the motions of the planets according to Copernicus's system.

Cornelius Van Drebble's Organ: Cornelius (van) Drebbel (1572–1633), Dutch inventor (of the submarine, among other things), built a water organ for James I. The other mechanical marvels mentioned in this sentence are now very obscure.

Lineal Arithmetic

William Playfair, 1798

Since the seventeenth century, scientists like Isaac Newton had been devising geometric constructions, of varying complexity, in which certain lines or areas would correspond to elapsed time while others corresponded to some force or distance moved. But William Playfair (brother of John Playfair, who was well known as a mathematician, geologist, and translator of Euclid) was practically the first to make graphs which plotted observed, as opposed to calculated, quantities against time; here he uses them to display the progress of international trade. For Playfair this was an application of "the principles of geometry to matters of finance," and one which he felt he needed to defend robustly against criticism.

William Playfair (1759–1823), *Lineal Arithmetic; Applied to show the Progress of the Commerce and Revenue of England During the Present Century; which is Represented and Illustrated by thirty-Three Copper-Plate Charts. Being a Useful Companion for the Cabinet and Counting house.* (London, 1798), pp. 5–11.

As the knowledge of mankind increases, and transactions multiply, it becomes more and more desirable to *abbreviate* and *facilitate* the modes of conveying information from one person to another, and from one individual to the many.

Algebra has abbreviated Arithmetical Calculations; Logarithmic Tables have shortened and simplified questions in Geometry. The study of History, Genealogy, and Chronology has been much improved by copperplate Charts; and it is now thirteen years since I first thought of applying lines to subjects of Finance.

At the time when this invention made its first appearance it was much approved of in England; Mr. Corry° applied the same mode to the Finances of Ireland, and my original work was translated and engraved in France, two years after, when it met with much approbation and success.

I confess I was very anxious to find out whether I was actually the first who applied the principles of geometry to matters of finance, as it had long before been applied to chronology with great success. I am now satisfied, upon due enquiry, that I was the first, for during eleven years I have never been able to learn that any thing of a similar nature had ever before been produced.

To those who have studied geography, or any branch of mathematics, these Charts will be perfectly intelligible. To such, however, as have not, a short explanation may be necessary.

The advantage proposed by those Charts (see Figure 10.2) is not that of giving a more accurate statement than by figures, but it is to give a more simple and permanent idea of the gradual progress and comparative amounts, at different periods, by presenting to the eye a figure, the proportions of which correspond with the amount of the sums intended to be expressed.

As the eye is the best judge of proportion, being able to estimate it with more quickness and accuracy than any other of our organs, it follows that wherever *relative quantities* are in question, a gradual increase or decrease of any revenue, receipt or expenditure of money, or other value, is to be stated, this mode of representing it is peculiarly applicable; it gives a simple, accurate, and permanent idea, by giving form and shape to a number of separate ideas, which are otherwise abstract and unconnected. In a numerical table there are as many distinct ideas given, and to be remembered, as there are sums; the order and progression, therefore, of those sums are also to be recollected by *another* effort of memory; while this mode unites *proportion, progression,* and *quantity* all under one simple impression of vision, and consequently one act of memory.

This method has struck several persons as being fallacious, because geometrical measurement has not any relation to money or to time, yet here it is made to represent both. The most familiar and simple answer to this objection is by giving an example. Suppose the money received by a man in trade were all in guineas, and that every evening he made a single pile of all the guineas received during the day. Each pile would represent a day, and its height would be proportioned to the receipts of that day, so that by this plain operation *time, proportion,* and *amount* would all be physically combined.

Lineal arithmetic then, it may be averred, is nothing more than those piles of guineas represented on paper, and on a small scale, in which

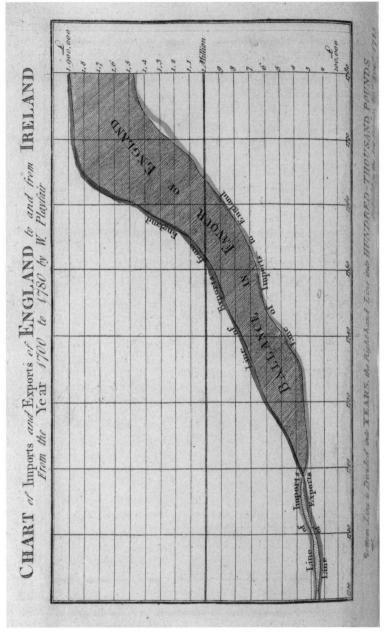

Figure 10.2. A specimen of "lineal arithmetic." (Playfair, facing p. 8. ©The British Library Board. 8503.ee.18.)

an inch (perhaps) represents the thickness of five millions of guineas, as in geography it does the breadth of a river, or any other extent of country.

My reason for adopting this mode of stating the present revenue of the nation is for the purpose of comparing it with the past, as also of comparing the progress of the revenues of the state with the progress of the influx of wealth from other countries; for it is not from the present state of things, uncompared with the past, that any conclusion can be drawn.

The human mind has been so worked upon for a number of years past, and the same subjects have been so frequently brought forward, that it is necessary to produce novelty, but above all to aim at facility, in communicating information, for the desire of obtaining it has diminished in proportion as disgust and satiety have increased.

⌊…⌋ I have succeeded in proposing and putting in practice a new and useful mode of stating accounts ⌊…⌋; as much information may be *obtained in five minutes as would require whole days to imprint on the memory in a lasting manner by a table of figures.*

As to the materials, they are taken from the accounts laid every year before the House of Commons, therefore may be depended upon as the best that are to be procured.

Note

Mr. Corry: James Corry (dates unknown) had contributed charts of the revenue and debts of Ireland to Playfair's *Commercial and Political Atlas* in 1786.

Astronomy in New South Wales

Charles Stargard Rumker, 1825

Charles Stargard Rumker (who was born in Germany and used several other forms of his name), astronomer, won fame in his field for the rediscovery of a lost comet (Encke's comet) in 1822. He worked at the Parramatta Observatory in Australia, and at the Hamburg Observatory in Germany. The passage given below sets out some of the purely astronomical (or indeed geometrical) reasons why an observatory in the southern hemisphere was desirable. Coming from an era and a milieu in which it was more usual to speak of knowledge being exported to the (passively grateful) colonies, it is a rare acknowledgment of the role of new geographical territories (not, alas, their indigenous cultures) in producing knowledge that was otherwise unavailable.

Charles Stargard Rumker (1788–1862), "On the Astronomy of the Southern Hemisphere (Read 13th March, 1822, before the Philosophical Society of Australia)" in Barron Field, *Geographical Memoirs on New South Wales; by various hands: containing An Account of the Surveyor General's late Expedition to two new Ports; The Discovery of Moreton Bay River, with the Adventures for seven months there of two shipwrecked Men; A Route from Bathurst to Liverpool Plains: together with other papers on the Aborigines, the Geology, the Botany, the Timber, the Astronomy, and the Meteorology of New South Wales and Van Diemen's Land* (London, 1825), pp. 260–261, 266–268.

The advantage of an observatory in the southern hemisphere is obvious. Merely for the purpose of observing the transit of Venus over the sun's disk, expensive expeditions were sent to different parts of the earth. By the English government, Captain Cook was sent to Otaheite, Mason and Dixon to the Cape of Good Hope, Dimond and Wales to Hudson's Bay. By the French government, the Abbe Chappe d'Autrouche was sent to Tobolsk in Siberia, and Pere Pingre to Rodriguez, one of the Mauritius Islands. The Spaniards sent Medina to California on the west coast of America. From Germany went Petre Hell to Wardehuus in Lapland; and many other expeditions were undertaken. We shall have this year the benefit of observing in New South Wales, on the 4th day of November, the transit of a planet over the sun's disc, without the inconvenience and expense of travelling.

A variety of interesting pursuits offer themselves in this yet-so-little-known part of the heavens. What a number of celestial bodies may, during centuries, have been roaming about in this wide field, that never rose to the arctic regions! Henceforth, none can escape. Sentinels being placed at the Cape of Good Hope and in Australia, no stranger can pass through the southern hemisphere without being hailed. Had these been on their posts thirty years ago, Enke's discovery of the periodical comet, which we expect to see this year, would not have been so long a secret, it being chiefly in high southern latitudes that it is visible. The comet that surprised us in 1819 below the north pole would have been seen long before in the southern hemisphere. The comet which we in Europe observed in February of the last year but for a short while, it absconding in the sun's rays, would, after its perihelian passage, have been seen by antarctic astronomers as bright as that of 1811.

To enumerate all the advantages which astronomy may derive from a fixed observatory in New South Wales would be an endless undertaking. To an astronomer, immediately after his departure from Europe, sailing down the Atlantic Ocean, with every step he takes in latitude, the heavens present a new scene. Thus, as he mounts over the earth's curved back, which till then interposed itself between him and the heavens' southern zones, he gradually sees down into a new field, richly sown with unknown stars, which to register and class is his pleasant duty. In this we follow, under much more favourable circumstances, a noble example first given in 1677 by Edmund Halley in St. Helena, but particularly by Nicolaus de la Caille, who (badly supplied by his government, and worse fitted out with instruments) formed, with indefatigable exertions, at the Cape of Good Hope from the year 1751 to 1754, a catalogue of 10,035 southern stars, inclosed within the tropic of Capricorn, determined the parallax of Mars and the moon, the length of the pendulum, measured a degree of the meridian, made magnetic observations, besides many other works, each of which would have immortalized him. But infinite is the task, and beyond human power. Neither Halley, La Caille, nor any of their successors, could or can complete it. Much is, and must be, left to do.

The Advantages of Mathematics

William Barnes, 1834

William Barnes was a poet and a philologist who worked for much of his life as a schoolmaster. His book on the advantages of mathematics was dedicated to Major-General Henry Shrapnel, "The Greatest Mathematician to whom the Author has had the Honor of Being Introduced" (and the inventor of the Shrapnel shell).

Barnes's remarkable, and remarkably optimistic, passage on the uses of mathematics and the usefulness of a mathematical education is representative of a trend that gathered momentum during the nineteenth century and will be seen again in our extracts from Sylvester and from Feynman.

W. Barnes (1801–1886), *A Few Words on the Advantages of a More Common Adoption of the Mathematics as a Branch of Education, or Subject of Study* (London, 1834), pp. 5–23.

Many persons unacquainted with the Mathematics are apt to conceive wrong ideas of their usefulness, thinking either that they are only applicable to such sciences as Astronomy and Navigation, and are therefore of too abstruse a character to be commonly useful, or that they consist only in the use of a few mathematical instruments, and the art of measuring a piece of ground, or finding a level, and are consequently of too narrow an application to be worth common cultivation; in both of which cases the truth that they are applicable and actually applied to almost all the arts of civilised society is overlooked.

The Mathematics then, it may be observed, consist of several branches, forming altogether the great science of lines, number, quantity, size, shape, distance, motion, and force, and every object in nature or art having either of those attributes is evidently susceptible, in some way or another, of mathematical investigation. These branches or sciences are of two kinds: one called pure or speculative Mathematics, and the other mixed or practical Mathematics; the former teaching general principles as un-applied (though applicable) to distinct objects, as Geometry, Algebra, and Arithmetic, and the latter consisting in the use of those general principles as applied to some branch of science or art, as Astronomy, Navigation, Mechanics, Problematical Geography, Hydrostatics, and Hydraulics, Optics, Perspective, and Architecture.

What! one may here exclaim, should one bewilder himself with all these sciences, none of which most likely would ever be of any use to him? No; surely not. I would only suggest that to Arithmetic (already duly appreciated) should be added the other branches of the pure Mathematics, Geometry (Euclid's Elements), and Algebra, the advantages of which I have now to show.

The elements of Geometry, as we have them in Euclid's system, form the basis of all the practical Mathematics, and are as a master-key that opens them all to our investigation and comprehension. They are a great store of principles which can be applied to hundreds of different objects, and from which hundreds of propositions can be derived as infallibly true as themselves. They are the alphabet by which we read the seemingly mysterious calculations of astronomers and engineers. He who is master of those principles is already in the ante-room to all the professions and crafts in which they are used, and by the opening of a single door (a little practical exemplification) he is easily introduced to either of them.

The elements of Geometry are applied even to the other branches of pure Mathematics, Arithmetic and Algebra: one of its propositions, for instance, giving us in the former, the Rule of Three; and in the latter, the conversion of a proportion into an equation.° Others again teach us to form the algebraical equations for curves, while Algebra, in its turn, investigates and proves for Arithmetic many of its rules; and as Geometry lends its principles to Algebra, so the latter helps the former by its operations, many geometrical problems being more easily solved by Algebra than by Geometry.

But Algebra is not to be taken up with Geometry because it assists it in its problems, but because it solves questions wholly beyond its power, and because the two sciences together form a complete system of human reasoning.

In our investigation of moral as well as mathematical truth, we reason by one of two methods: by synthesis, or analysis. That is, by establishing one truth as the consequence of another, and a third as that of the second, proceeding step by step, truth by truth, till at the end of the series we establish the object of our investigation as the necessary consequence of all the foregoing; or else by taking, as it were, the subject to pieces: separating from the truth sought all other things that are involved with it in the question, and bringing it out singly as the sole object of our search.

Now the elements of Geometry form a system of synthetical reasoning, as Algebra is one of analytical investigation; hence they are, as it were, necessary to each other, as one can be made to act where the other would fail. As a specimen of synthesis, Euclid's Elements are most admirable, Thales's famous proposition that "the square of the hypotenuse of a right-angled triangle is equal to the sum of the squares of the other two sides," being, for instance, the 47th of the first book: that is, the last in a series of 47 truths, and proved only as a consequence of all the others, each of which back to the second is a consequence of preceding ones, so that if a link of the chain were struck out, all beyond it would be lost. Algebra, again, is equally admirable as a system of analysis. The doctrine of simple equations is one of the most easy and least powerful of its branches, and yet, if we have a mass of gold and silver melted together, and weighing in air 106lb, and if we find by Hydrostatics that the mass weighs only 99lb. under water, and that pure gold loses 1/19th of its weight, and pure silver 1/10th of its weight under water, a simple equation founded on these data will readily analyse the mass (in calculation), and show how much gold and how much silver is in it.

Moral questions are more commonly investigated by synthetical reasoning; hence Geometry becomes most useful as a strengthener of the reason, a guide in our searches after moral truth, and a defence against the power of sophistry and false doctrine; of which our learned bodies are so convinced, that at the University of Oxford it is often taken as an equivalent for logic, and at Cambridge is an essential, as it was anciently in the philosophical school of Plato, "Let no one be admitted without Geometry" being one of his standing rules. To this use of Geometry Lord Byron alluded when he said of Madame de Staël "The want of a mathematical education, which might have served as a ballast to steady and help her into the port of reason, was always visible."—*Conversations of Lord Byron*, by the Countess of Blessington.

St. Paul gives us an interesting specimen of synthetical reasoning in the 15th chapter of his first Epistle to the Corinthians, from the 12th to the 20th verses.

That the study of the pure Mathematics qualifies one to some extent for any practical branch of the science has been shown by Smeaton,° who was originally (I believe) a Mathematical Instrument Maker in London, and afterwards the constructor of the Eddystone Lighthouse, because, having learnt and cultivated the Mathematics in his first profession, he became qualified as an Engineer to give the solution of a problem of force and resistance, and geometrical construction, so finely shown in that clever production of his mind.

This much being said of the connexion between the Mathematics and many of the arts of life, it may now be observed that they are well worth studying by those who may follow professions not connected with them, and even by those who may not have any need of following a profession at all, since, as has been shown before, they teach the mind a regular method of reasoning, and help us in our searches after moral truth. And there are few persons who do not often want to measure or reason upon areas and solids, solve geometrical problems and investigate principles of work, either to escape imposition, or calculate beforehand the expenses and likely results of conceived plans, whether in Building or Agriculture, in Geometry or Mechanics. Even, for example, if one wished to have a pit of a definite shape and size dug in a Hold, and a man wanted a fixed sum for excavating a cubic yard of soil, one might have in Geometry the means of finding beforehand the exact cost of it.

Without some knowledge of the Mathematics one cannot easily comprehend even many allusions, terms, observations, and formulae, often met with in scientific (even though not technical) books, nor comprehend, or at least not describe, a thousand things about us in Mechanics, Architecture, and the like. The sun's or moon's parallax, or the aberration of light, are words of little meaning without geometrical exemplifications, and, from a want of mathematical definitions, what periphrases must we sometimes use to describe something we have seen!—perhaps the frustra of a cone, the sector of a circle, or an octagonal pyramid. How we confound the different figures!—the ellipse with the oval, the parabolic curve with the elliptic, the sector of a circle with the segment, the cycloid with the circle, and though these things are of little importance as affecting one's well-being in life, everyone would wish to be correct in his conceptions, whether as a speaker or hearer.

And here I would observe to those who may object to the extension of popular education, that these few observations on the advantages of mathematical learning refer not to the poor, but to their betters. I am not recommending the use of Euclid's Elements in National Schools instead of the Bible, but I would recommend the pure Mathematics as a branch of study most highly useful to all who may follow or be likely to adopt a mathematical profession or trade, and as a most fitting branch of a genteel education for those who may not mean to take up a mathematical profession or need any profession at all.

Notes

the conversion of a proportion into an equation: a proportion is an equality of ratios; an equation is an equality of quantities.

Smeaton: John Smeaton (1724–1792), a Fellow of the Royal Society and the engineer of many major public building projects.

Sylvester Contra Huxley

J. J. Sylvester, 1870

This brief extract comes from the response to Thomas Henry Huxley by the well-known mathematician James Joseph Sylvester, president of the London Mathematical Society and later Savilian Professor of Geometry at Oxford. Huxley, "Darwin's bulldog," as the contemporary phrase had it, had cast aspersions on mathematics in a speech and in articles during 1868–1869, contrasting its methods unfavorably with those of inductive science and claiming that mathematics "is that which knows nothing of observation, nothing of experiment, nothing of induction, nothing of causation." Sylvester was not impressed. But there was, arguably, underlying agreement between the two men on the need for reform of the way mathematics and science were taught.

Published in the same volume were Sylvester's remarkable "laws of verse," partly in mathematical form and exemplifying that line of (particularly) nineteenth-century thinking which tended to see the potential for rational and quantitative systematization even in the most unlikely places.

J. J. Sylvester (1814–1897), *The Laws of Verse Or Principles of Versification Exemplified in Metrical Translations: together with an annotated reprint of The Inaugural Presidential Address to the Mathematical and Physical Section of the British Association at Exeter* (London, 1870), pp. 122–123.

Some people have been found to regard all mathematics, after the 47th proposition of the first book of Euclid, as a sort of morbid secretion, to be compared only with the pearl said to be generated in the diseased oyster, or, as I have heard it described, "une excroissance maladive de l'esprit humain."° Others find its justification, its "raison d'être," in its being either the torch-bearer leading the way, or the handmaiden holding up the train of Physical Science; and a very clever writer in a recent magazine article expresses his doubts whether it is, in itself, a more serious pursuit, or more worthy of interesting an intellectual human being, than the study of chess problems or Chinese puzzles. What is it to us, they say, if the three angles of a triangle are equal to two right angles, or if every even number is, or may be, the sum of two primes, or if every equation of an odd degree must have a real root? How dull, stale, flat, and unprofitable are such and such like announcements! Much more interesting to read an account of a marriage in high life, or the details of an international boat-race. But this is like judging of architecture from being shown some of the brick and mortar, or even a quarried stone, of a public building, or of painting from the colours mixed on the palette, or of music by listening to the thin and screechy sounds produced by a bow passed haphazard over the strings of a violin. The world of ideas which it discloses or illuminates, the contemplation of divine beauty and order which it induces, the harmonious connection of its parts, the infinite hierarchy and absolute evidence of the truths with which it is concerned, these, and such like, are the purest grounds of the title of mathematics to human regard, and would remain unimpeached and unimpaired were the plan of the universe unrolled like a map at our feet, and the mind of man qualified to take in the whole scheme of creation at a glance.

Note

une excroissance maladive de l'esprit humain: a diseased excrescence of the human spirit.

What a Mathematical Proposition Is

Cassius Jackson Keyser, 1929

Cassius Jackson Keyser, who spent much of his career at Columbia University, published a number of semipopular books and pamphlets taking a broadly philosophical approach to mathematics. Introducing *The Pastures of Wonder*, he wrote that it was "not a 'Story' nor a 'Romance' nor a 'jazzy' attempt at popularization"; the defensive tone would become a common one later in the twentieth century when talking about mathematics. Insistent that neither jargon, nor geometrical method, is the point, but picking up a theme we have seen earlier in this chapter, he makes logic an essential, even a defining, part of mathematics.

Cassius Jackson Keyser (1862–1947), *The Pastures of Wonder: The Realm of Mathematics and the Realm of Science* (New York, 1929), pp. 39–45. Copyright © (1929) Columbia University Press. Reprinted with permission of the publisher.

What is a mathematical proposition? The answer is: A mathematical proposition is a hypothetical proposition that is regarded by the mathematical world as having been demonstrated. In other words, it is a hypothetical proposition whose conclusion or implicate, q, is regarded by the competent as having been logically deduced from the proposition's hypothesis, or implier, p. "Mathematics," says Pieri° "is the hypothetico-deductive science." I venture to believe that we are now beginning to see what he meant. I have said "beginning to see," for the *full* significance of his *mot* is too profound, too subtle and too vast to be so quickly disclosed, and, for an adequate understanding of it, our meditation has a fairly long course yet to run.

Two myth-destroying facts and their sigificance

The preceding paragraph contains two definitions of major importance: a definition of the term, Mathematical Proposition, and a definition of Mathematics. There are two facts about them which we must not fail to observe, for the facts in question are fatal, or ought to be fatal, to a pair of ages-old and still reigning myths regarding the essential nature and the scope of the mathematical method. One of the facts is that neither the definition of mathematical proposition nor that of mathematics says anything about quantities or about numbers or about geometric entities or about any other specific kind of subject-matter. The other fact is that neither of the two definitions says anything about those strange, repellent, world-frightening signs and symbols which increasingly abound in mathematical literature and which are, commonly, about the only things of which the word mathematics recalls even so much as a vague and jumbled impression. The critical significance of the two facts is fundamental. Let us examine it somewhat attentively.

Sheer mathematics is form without content

The first one of the mentioned facts signifies that, when thinking mathematically, we need not be thinking about quantities or magnitudes or about numbers or about geometric entities or spatial configurations or about any other specific kind of subject-matter; what is much more, it signifies that, when thinking mathematically, we are thinking in a way which, because it is independent of what is peculiar to any kind of subject-matter, is *applicable* to *all* kinds—available, that is, in every field of thought. The thesis just stated regarding the general availability of mathematical thinking is so important for the prosperous conduct of human life that its importance cannot be exaggerated. ⌐...⌐ For the present I will merely exemplify it by a simple example familiar to all.

Consider the hypothetical proposition: If John Doe was in Chicago at midnight of June 30, 1926, then he did not at that time stab Richard Roe in New York City. Ordinarily the proposition would be regarded as obviously true. Yet, strictly taken, it is not true, for the conclusion cannot be deduced from the stated hypothesis. The deduction becomes possible if and only if the stated hypothesis be enlarged by adding to it certain propositions which the defendant's counsel might think it unnecessary to state explicitly because a juror would unconsciously take them for granted. I mean such propositions as that the alleged stabbing required the presence of the stabber at the time and place of the deed and that the two cities mentioned are such that Doe could not have been in both of them at the same time, which of course he might have been were the cities overlapping. If the stated hypothesis be thus rightly enlarged, the deduction in question becomes possible and the proposition true and logically demonstrable. It is thus evident that every *alibi* defense involves the application of a genuine bit of mathematics, a genuine bit of hypothetico-deductive thinking. By a little observation and reflection readers can discover for themselves that many similar examples occur, in more or less disguised and often imperfect form, here, there and yonder, in all connections and situations, high or low, near or remote, where human beings have tried to *infer*.

The essence of mathematics is not in its symbols

I have already drawn attention to the fact that the definition of a mathematical proposition and the definition of mathematics are both of them silent respecting those peculiar signs and symbols which professional mathematicians so much employ and without which, it is commonly believed, mathematical thinking would be impossible. What does that silence signify? It signifies that the mentioned belief is a myth. Mathematical signs and symbols are nothing but linguistic devices gradually invented for the purpose of economising intellectual energy and, because they serve that purpose so well, their use is highly expedient. But a vast deal of mathematical thinking was done before they were invented and much of it is now done without their use, by means of the words or symbols of ordinary speech, just as agriculture existed for ages before the invention of modern agricultural machinery and is even now extensively carried on without the use of such machinery, by means of primitive implements. For mathematical purposes ordinary words are primitive instruments. The economic power of mathematical symbols is indeed very great, so great that mathematicians have been thereby enabled to construct many a doctrine that they could not have constructed without using them. And though such a doctrine, once it has been thus constructed, could by great labor be translated into ordinary language, yet the resulting discourse would be so prolix, involved, and cumbrous that none but a god could read it understandingly and no god would do it unless he were a divine fool. Notwithstanding the immense service rendered by the symbols in question, it is no more true to say that without them there could be no mathematical thinking than to say that without the modern means of passenger transportation there could be no travelling or that fighting would be impossible were there no modern instruments of war.

Note

Pieri: Mario Pieri (1860–1913), Italian geometer and philosopher of mathematics.

The Character of Physical Law

Richard P. Feynman, 1965

Richard Feynman was one of the great physicists of the twentieth century and one of its greatest scientific teachers and popularizers; these extracts on the (mathematical) character of physical laws display his lucid and charismatic style of exposition. They also exemplify something close to the high-water mark of optimism about the achievements and potential of a mathematical world view.

The Relation of Mathematics to Physics

In thinking out the applications of mathematics and physics, it is perfectly natural that the mathematics will be useful when large numbers are involved in complex situations. In biology, for example, the action of a virus on a bacterium is unmathematical. If you watch it under a microscope, a jiggling little virus finds some spot on the odd shaped bacterium—they are all different shapes—and maybe it pushes its DNA in and maybe it does not. Yet if we do the experiment with millions and millions of bacteria and viruses, then we can learn a great deal about the viruses by taking averages. We can use mathematics in the averaging, to see whether the viruses develop in the bacteria, what new strains and what percentage, and so we can study the genetics, the mutations and so forth.

To take another more trivial example, imagine an enormous board, a chequerboard to play chequers or draughts. The actual operation of any one step is not mathematical—or it is very simple in its mathematics. But you could imagine that on an enormous board, with lots and lots of pieces, some analysis of the best moves, or the good moves or bad moves, might be made by a deep kind of reasoning which would involve somebody having gone off first and thought about it in great depth. That then becomes mathematics, involving abstract reasoning. Another example is switching in computers. If you have one switch, which is either on or off, there is nothing very mathematical about that, although mathematicians like to start there with their mathematics. But with all the interconnections and wires, to figure out what a very large system will do requires mathematics.

I would like to say immediately that mathematics has a tremendous application in physics in the discussion of the detailed phenomena in complicated situations, granting the fundamental rules of the game. That is something which I would spend most of my time discussing if I were talking only about the relation of mathematics and physics. But since this is part of a series of lectures on the character of physical law I do not have time to discuss what happens in complicated situations, but will go immediately to another question, which is the character of the fundamental laws.

If we go back to our chequer game, the fundamental laws are the rules by which the chequers move. Mathematics may be applied in the complex situation to figure out what in given circumstances is a good move to make. But very little mathematics is needed for the simple fundamental character of the basic laws. They can be simply stated in English for chequers.

The strange thing about physics is that for the fundamental laws we still need mathematics. I will give two examples, one in which we really do not, and one in which we do. First, there is a law in physics called Faraday's law, which says that in electrolysis the amount of material which is deposited is proportional to the current and to the time that the current is acting. That means that the amount of material deposited is proportional to the charge which goes through the system. It sounds very mathematical, but what is actually happening is that the electrons going through the wire each carry one charge. To take a particular example, maybe to deposit one atom requires one electron to come, so the number of atoms that are deposited is necessarily equal to the number of electrons that flow, and thus proportional to the charge that goes through the wire. So that mathematically-appearing law has as its basis nothing very deep, requiring no real knowledge of mathematics. That one electron is needed for each atom in order for it to deposit itself is mathematics, I suppose, but it is not the kind of mathematics that I am talking about here.

On the other hand, take Newton's law for gravitation, which ⌐...⌐ I discussed last time. I gave you the equation:

$$F = G\frac{mm'}{r^2}$$

just to impress you with the speed with which mathematical symbols can convey information. I said that the force was proportional to the product of the masses of two objects, and inversely as the square of the distance between them, and also that bodies react to forces by changing their speeds, or changing their motions, in the direction of the force by amounts proportional to the force and inversely proportional to their masses. Those are words all right, and I did not necessarily have to write the equation. Nevertheless it is kind of mathematical, and we wonder how this can be a fundamental law. What does the planet do? does it look at the sun, see how far away it is, and decide to calculate on its internal adding machine the inverse of the square of the distance, which tells it how much to move? This is certainly no explanation of the machinery of gravitation! You might want to look further, and various people have tried to look further. Newton was originally asked about his theory—"But it doesn't mean anything—it doesn't tell us anything." He said, "It tells you *how* it moves. That should be enough. I have told you how it moves, not why."

Up to today, from the time of Newton, no one has invented ₍a₎ theoretical description of the mathematical machinery behind this law which does not either say the same thing over again, or make the mathematics harder, or predict some wrong phenomena. So there is no model of the theory of gravitation today, other than the mathematical form.

If this were the only law of this character it would be interesting and rather annoying. But what turns out to be true is that the more we investigate, the more laws we find, and the deeper we penetrate nature, the more this disease persists. Every one of our laws is a purely mathematical statement in rather complex and abstruse mathematics. Newton's statement of the law of gravitation is relatively simple mathematics. It gets more and more abstruse and more and more difficult as we go on. Why? I have not the slightest idea. It is only my purpose here to tell you about this fact. The burden of the lecture is to emphasize the fact that it is impossible to explain honestly the beauties of the laws of nature in a way that people can feel, without their having some deep understanding of mathematics. I am sorry, but this seems to be the case.

Up to today, from the time of Newton, no one has invented ₍a₎ theoretical description of the mathematical machinery behind this law which does not

It always bothers me that, according to the laws as we understand them today, it takes a computing machine an infinite number of logical operations to figure out what goes on in no matter how tiny a region of space, and no matter how tiny a region of time. How can all that be going on in that tiny space? Why should it take an infinite amount of logic to figure out what one tiny piece of space/time is going to do? So I have often made the hypothesis that ultimately physics will not require a mathematical statement, that in the end the machinery will be revealed, and the laws will turn out to be simple, like the chequer board with all its apparent complexities. But this speculation is of the same nature as those other people make—"I like it," "I don't like it"—and it is not good to be too prejudiced about these things.

To summarize, I would use the words of Jeans, who said that "the Great Architect seems to be a mathematician." To those who do not know mathematics it is difficult to get across a real feeling as to the beauty, the deepest beauty, of nature. C.P. Snow talked about two cultures.° I really think that those two cultures separate people who have and people who have not had this experience of understanding mathematics well enough to appreciate nature once.

It is too bad that it has to be mathematics, and that mathematics is hard for some people. It is reputed—I do not not know if it is true—that when one of the kings was trying to learn geometry he complained that it was difficult. And Euclid said, "There is no royal road to geometry." And there *is* no royal road. Physicists cannot make a conversion to any other language. If you want to learn about nature, to appreciate nature, it is necessary to understand the language that she speaks in. She offers her information only in one form; we are not so unhumble as to demand that she change before we pay any attention.

Note

C.P. *Snow talked about two cultures*: Charles Percy, Baron Snow (1905–1980), novelist, essayist, and civil servant who famously criticized the modern gulf between the arts and the sciences in writings and lectures in the 1950s, describing the situation in the still-current phrase "the two cultures."

Our Invisible Culture

Allen L. Hammond, 1978

Lynn Arthur Steen is an Emeritus Professor of Mathematics at St. Olaf College, Northfield, Minnesota, and has worked across a wide range of curriculum design and reform issues in undergraduate and school mathematics. His edited collection, *Mathematics Today*, contains a marvelous range of discussions of what mathematics is and does; here we see a passage by Allen L. Hammond, better known as a writer on economics and energy policy, discussing what mathematics is and how and by whom it is done. It represents a very different approach to these questions from Keyser's. That mathematics has a human face hardly needs emphasizing in this book, but it is interesting to see Hammond suggest that that face is elusive. Concerning his question about mathematics in the newspapers, see "The Monster Unveiled," in Chapter 5.

An inquiry into mathematics and mathematicians might begin with certain curious facts. One is that mathematics is no longer an especially uncommon pursuit. Never mind that a multitude of mathematicians seems a contradiction in terms. The universities are simply teeming with them. The latest figures compiled by the National Science Foundation show that there are as many mathematicians in the United States as there are physicists or economists. Mathematicians are not a rare breed, simply an invisible one. It is a multitude singularly accomplished at keeping out of the public eye. Who has ever seen a mathematician on television, or read of their exploits in the newspapers?

A second fact about this reticent profession is even more startling. All those people are busy doing something, including some very remarkable somethings. In all its long history extending back 25 centuries, mathematics has never been more vigorous, more active than now. Within this century mathematicians have experienced philosophical upheavals and intellectual advances as profound as those that have catapulted physicists into fame or transformed economists into the indispensible advisors of governments. The foundations of mathematics itself have been challenged and rewritten, whole new branches have budded and flourished, seemingly arcane bits of theory have become the dicta for giant industries. Yet this drama has been played out in near obscurity. The physical concepts of relativity and subatomic particles have entered the language, the gross national product is reported to millions of living rooms, but it is as if the very texture of mathematics is antithetical to broad exposure. What is it in the nature of this unique field of knowledge, this unique human activity that renders it so remote and its practitioners so isolated from popular culture?

In searching for a foothold to grapple with this elusive subject, an Inquirer is struck by the contradictions that abound. For example, mathematics is nearly always described as a branch of science, the essence of pure reason. Beyond doubt mathematics has proved to be profoundly useful, perhaps even essential, to the modern edifice of science and its technological harvest. But mathematicians persist in talking about their field in terms of an art—beauty, elegance, simplicity—and draw analogies to painting, music. And many mathematicians would heatedly deny that their work is intended to be useful, that it is in any sense motivated by the prospect of practical application. A curious usefulness, an aesthetic principle of action; it is a dichotomy that will bear no little scrutiny in what is to come.

A further contradiction arises from the stuff of mathematics itself. It is in principle not foreign to our experience, since the root concepts are those of number and of space, intuitively familiar even to the child who asks "how many" or "how large." But the axiomatization and elaboration of these concepts has gone quite far from these simple origins. The abstraction of number to quantitative relationship of all kinds, the generalization of distance and area first to idealized geometrical figures and then to pure spatial forms of diverse types are large steps. Somewhere along the lengthy chains of logic that link modern mathematics to more primitive notions, a transmutation has occurred—or so it often seems to outsiders— and we can no longer recognize the newest branches on the tree of mathematics as genetically related to the roots. The connection is obscured, the terminology baffling. Is any of it for real? Do these abstractions and elaborations genuinely expand our understanding of number and of space, or do they amount to an empty house of theorems?

Mathematicians bristle at such questions. But it is not surprising that there is a popular tendency to dismiss much of this unfamiliar stuff as the subtle inventions of clever minds and having no important relationship to reality. What is surprising is that mathematicians do not agree among themselves whether mathematics is invented or discovered, whether such a thing as mathematical reality exists or is illusory. Is the tree of mathematics unique? Would any intelligence (even a nonhuman one) build similar structures of logic? How arbitrary is the whole of mathematical knowledge? These two are points worth additional inquiry.

We might also learn something of the end result, however incomprehensible, if we could see the process by which it is made and know more of the makers. Should we pity the poor mathematician, condemned to serve his or her days bound to a heavy chain of cold logic? How does that image jibe with the white-hot flashes of insight, the creative "highs," so often reported, or the intensely human character of mathematicians in the flesh? Clearly, a suitable subject for this inquiry is the nature of mathematicians themselves, their motivations, their trials, their rewards, and how they spend their days.

A final question might be directed toward the place of mathematics in our culture. There are those, including Plato, who have identified mathematics with the highest ideal of civilization—a lofty claim indeed. A claim more often made and subscribed to by mathematiciains is that mathematics is one of the finest flowerings of the human spirit, a cathedral of enduring knowledge built piece by piece over the ages. But if so it is a cathedral with few worshippers, unknown to most of humankind. Mathematics plays no role in mass culture, it cannot claim to evoke the sensibilities and inspire the awe that music and sculpture do, it is not a significant companion in the lives of more than a very few. And yet it is worth asking whether mathematics is essentially remote, or merely poorly communicated. Perhaps it is a remediable ignorance, not an inability, that now limits appreciation and enjoyment of mathematical intuitions by a wider audience; perhaps our culture is only reaching the stage at which mathematics can begin to penetrate a larger consciousness.

11

The Mathematicians Who Never Were: Fiction and Humor

Let me see: four times five is twelve, and four times six is thirteen, and four times seven is—oh dear! I shall never get to twenty at that rate! However, the Multiplication Table don't signify.
—Lewis Carroll, *Alice's Adventures in Wonderland*, 1865, chapter 2.

DOES IT SIGNIFY? FULL OF TERMS THAT OUTSIDERS DON'T UNDERSTAND, FULL of things that don't exist in the real world (lines with no breadth, points with no size), full of apparent nonsense yet full of ambition to change the world: mathematics has long been a rich field for fiction, humour and satire. Fictional treatments of mathematics often focus on individual mathematicians, and these "mathematicians who never were" are much on display in this chapter. They include desert-island autodidacts, astronomers both admirable and absurd, and fantastic visions of the mathematicians of the distant past or of other worlds.

This riot of characters, and the other reflections on mathematics chosen for this chapter, articulate some serious concerns, too. Could mathematics threaten the stability of society? Could it distract us from the real business of life? Damage our health? Our minds? A reverse side to the last chapter's optimism about mathematics is on view here, and we hear the voices of individuals who were genuinely concerned about what mathematics would and could do to the world.

But it's not all gloom. These fictions show the capacity of mathematics to ennoble, to intrigue, to liberate, and of course to entertain. And, finally, they allow us to wonder: what might mathematics and mathematicians be like in the future?

If, for example, a Selenite is destined to be a mathematician ⌊...⌋ his brain grows continually larger, at least so far as the portions engaging in mathematics are concerned; they bulge ever larger and seem to suck all life and vigour from the rest of his frame. His limbs shrivel, his heart and digestive organs diminish, his insect face is hidden under its bulging contours. His voice becomes a mere stridulation for the stating of formula; he seems deaf to all but properly enunciated problems. The faculty of laughter, save for the sudden discovery of some paradox, is lost to him; his deepest emotion is the evolution of a novel computation. (H.G. Wells, *The First Men in the Moon* (London, 1901), Chapter 24.)

For some different possibilities, read on.

Spider-Men and Lice-Men

Margaret Cavendish, 1666

Margaret Cavendish, Duchess of Newcastle, a versatile writer and the only woman to attend a meeting of the Royal Society in the seventeenth century, included the description of a "Blazing-World" in her *Observations upon Experimental Philosophy* (1666). In it she cast herself as the empress of an imaginary world and poked fun at various learned groups under the guise of "ant-men," "lice-men," "bear-men," and so on. The mathematical and scientific experts come off particularly badly, with the experimental philosophers ordered to break their telescopes because they cause so much trouble, the mathematicians told that their writings are incomprehensible, and the society of geometers "dissolved" because they have involved themselves too much in impossible scientific tasks.

Margaret Cavendish (1623–1673), *The Description of a new World, called The Blazing-World. Written By the Thrice Noble, Illustrious, and Excellent Princesse, the Duchess of Newcastle* (London: 2nd edition, 1688), pp. 15–16, 25–29, 55–56.

The rest of the Inhabitants of that World were men of several different sorts, shapes, figures, dispositions, and humors ⌊...⌋; some were Bear-men, some Worm-men, some Fish- or Mer-men, otherwise called Sirens; some Bird-men, some Fly-men, some Ant-men, some Geese-men, some Spider-men, some Lice-men, some Fox-men, some Ape-men, some Jackdaw-men, some Magpie-men, some Parrot-men, some Satyrs, some Giants, and many more, which I cannot all remember. And of these several sorts of men,

Figure 11.1. Margaret Cavendish, inventor and empress of the "Blazing-World." (Margaret Cavendish, *Playes* (1662), frontispiece. © Princeton University Library. Rare Books Division, Department of Rare Books and Special Collections, 17th-752.)

each followed such a profession as was most proper for the nature of their species, which the Empress encouraged them in, especially those that had applied themselves to the study of several Arts and Sciences, for they were as ingenious and witty in the invention of profitable and useful Arts as we are in our world—nay, more—and to that end she erected Schools, and founded several Societies. The Bear-men were to be her Experimental Philosophers, the Bird-men her Astronomers, the Fly-, Worm- and Fish-men her Natural Philosophers, the Ape-men her Chymists, the Satyrs her Galenic Physicians, the Fox-men her Politicians, the Spider- and Lice-men her Mathematicians, the Jackdaw-, Magpie- and Parrot-men her Orators and Logicians, the Giants her Architects, etc.

⌀

To avoid hereafter tedious disputes, and have the truth of the Phenomena of Celestial Bodies more exactly known, ꜣthe Empressꜣ commanded the Bear-men, which were her Experimental Philosophers, to observe them through such Instruments as are called Telescopes, which they did according to her Majesty's Command. But these Telescopes caused more differences and divisions amongst them than ever they had before: for some said they perceived that the Sun stood still, and the Earth did move about it, others were of opinion that they both did move, and others said, again, that the Earth stood still, and the Sun did move. Some counted more Stars than others; some discovered new Stars never seen before; some fell into a great dispute with others concerning the bigness of the Stars; some said the Moon was another World like their Terrestrial Globe, and the spots therein were Hills and Valleys, but others would have the spots to be the Terrestrial parts, and the smooth and glossy parts the Sea.

At last, the Empress commanded them to go with their Telescopes to the very end of the Pole that was joined to the World she came from, and try whether they could perceive any Stars in it; which they did, and, being returned to her Majesty, reported that they had seen three Blazing-Stars appear there, one after another in a short time, whereof two were bright, and one dim. But they could not agree neither in this observation: for some said, It was but one Star which appeared at three several times, in several places; and others would have them to be three several Stars, for they thought it impossible that those three several appearances should have been but one Star, because every Star did rise at a certain time, and appeared in a certain place, and did disappear in the same place. Next, It is altogether improbable, said they, That one Star should fly from place to place, especially at such a vast distance, without a visible motion, in so short a time, and appear in such different places, whereof two were quite opposite, and the third side-ways. Lastly, If it had been but one Star, said they, it would always have kept the same splendor, which it did not, for, as above mentioned, two were bright, and one was dim.

After they had thus argued, the Empress began to grow angry at their Telescopes, that they could give no better Intelligence; for, said she, now I do plainly perceive that your Glasses are false Informers, and instead of discovering the Truth, delude your Senses. Wherefore I Command you to break them, and let the Bird-men trust only to their natural eyes, and examine Celestial Objects by the motions of their own Sense and Reason.

The Bear-men replied that it was not the fault of their Glasses, which caused such differences in their Opinions, but the sensitive motions in their Optic organs did not move alike, nor were their rational judgments always regular.

To which the Empress answered that if their Glasses were true Informers, they would rectify their irregular Sense and Reason. But, said she, Nature has made your Sense and Reason more regular than Art has your Glasses; for they are mere deluders, and will never lead you to the knowledge of Truth. Wherefore I command you again to break them; for you may observe the progressive motions of Celestial Bodies with your natural eyes better then through Artificial Glasses.

The Bear-men, being exceedingly troubled at her Majesty's displeasure concerning their Telescopes, kneeled down, and in the humblest manner petitioned that they might not be broken; for, said they, we take more delight in Artificial delusions than in Natural truths. Besides, we shall want Employments for our Senses, and Subjects for Arguments; for, were there nothing but truth, and no falsehood, there would be no occasion to dispute, and by this means we should want the aim and pleasure of our endeavours in confuting and contradicting each other; neither would one man be thought wiser than another, but all would either be alike knowing and wise, or all would be fools. Wherefore we most humbly beseech your Imperial Majesty to spare our Glasses, which are our only delight, and as dear to us as our lives.

The Empress at last consented to their request, but upon condition that their disputes and quarrels should remain within their Schools, and cause no factions or disturbances in State, or Government. The Bear-men, full of joy, returned their most humble thanks to the Empress, and to make her amends for the displeasure which their Telescopes had occasioned, told her Majesty that they had several other artificial Optic-Glasses, which they were sure would give her Majesty a great deal more satisfaction. Amongst the rest, they brought forth several Microscopes, by the means of which they could enlarge the shapes of little bodies, and make a Louse appear as big as an Elephant, and a Mite as big as a Whale.

⚬

The Empress having hitherto spent her time in the Examination of the Bird-, Fish-, Worm-, and Ape-men, etc., and received several Intelligences from their several employments, at last had a mind to divert herself after her serious discourses, and therefore she sent for the Spider-men, which were her Mathematicians, the Lice-men, which were her Geometricians, and the Magpie-, Parrot- and Jackdaw-men, which were her Orators and Logicians.

The Spider-men came first, and presented her Majesty with a table full of Mathematical points, lines and figures of all sorts of squares, circles, triangles, and the like, which the Empress, notwithstanding that she had a very ready wit, and quick apprehension, could not understand, but the more she endeavoured to learn, the more was she confounded. Whether they did ever square the circle, I cannot exactly tell, nor whether they could make imaginary points and lines, but this I dare say: That their points and lines were so slender, small and thin, that they seemed next to Imaginary. The Mathematicians were in great esteem with the Empress, as being not only the chief Tutors and Instructors in many Arts, but some of them excellent Magicians and Informers of Spirits, which was the reason their Characters were so abstruse and intricate that the Empress knew not what to make of them. There is so much to learn in your Art, said she, that I can neither spare time from other affairs to busy myself in your profession, nor, if I could, do I think I should ever be able to understand your Imaginary points, lines and figures, because they are Non-beings.

Then came the Lice-men, and endeavoured to measure all things to a hair's breadth, and weigh them to an Atom; but their weights would seldom agree, especially in the weighing of Air, which they found a task impossible to be done; at which the Empress began to be displeased, and told them, that there was neither Truth nor Justice in their Profession, and so dissolved their society.

In the Court of Lilliput

"Captain Gulliver," 1727

Not the work of Jonathan Swift, but that of an imitator, possibly Mrs. Eliza Heywood (1693?–1756), author and actress, this utopian fantasy displays an imagined attitude toward mathematics even more severe than that in Margaret Cavendish's Blazing-World and illustrates vividly the notion that mathematics could be politically and even morally dangerous. We have met several times in this volume the notion that mathematics might be mistaken for magic or conjuring, that it might cause moral or social disturbance or merely be a dangerous misuse of resources; but this is the only occasion on which we see it, for any of those reasons, actually outlawed.

Gulliver, Captain, *Memoirs of the court of Lilliput. Written by Captain Gulliver.* (London, 1727), pp. 121–122, 126–127, 128–132, 134–135.

At the farther end of my Apartment I saw some little confused Spots and Lines drawn athwart each other in a Mathematical manner, which, though I had lived here for many Months, I had never observed before ⌞...⌟.

The Figures I saw ⌞...⌟ as I approached increased my Wonder. I never saw in England a Globe more exactly drawn; I do not believe the nicest Mathematician could have found fault with the smallest Line. How, said I to myself, can these People have so just a notion of the Position of the World, yet imagine there are no parts of it habitable but that they possess, and the small Island of Blefuscu?

❧

⌞I⌟ should have been tempted to have believed that Painting had been done by an European Hand, if the size of the Temple in which it was had not convinced me it could only be of service to a Lilliputian Race.

⌞...Keldresal⌟ told me that had I been a Lilliputian born, or had lived among them long enough to be acquainted with their Laws, to have brought those Figures to light would have drawn on me some very severe

Punishment, but as I was a Stranger, and had been guilty only through Ignorance, 'twas probable I might obtain Pardon from the Emperor, if he should happen to know it. However, he advised me to conceal what I had done, and erase the Figures, if by any means I could, so as they might not be seen by any that came to visit me.

I begged he would inform me of the Reasons which made me guilty, if he could possibly do it without becoming so himself.

This place, said he, which is now allotted for your Apartment, was formerly a Temple, and the most magnificent one in the whole Kingdom; this Painting that you have so miraculously discovered was done by the greatest Artist of his Time, from a Draught given him by the first and perhaps the greatest Philosopher, Mathematician, and Geographer that ever the World produced. He divided the Globe by straight and oblique Lines, in the manner you see it here deciphered, foretold the Change of Weather, counted the number of the Stars, and prefixed certain Times for the rising and setting of the Sun in such and such Seasons of the Year. He was greatly applauded for the success of his Labours, and our History informs us that never Man received more substantial Proofs of Esteem and Admiration. This brought the Art or Study of Mathematics so much in fashion, that all our young Nobility and Gentry bent their Minds this way, and presently after rose up a number of imaginary Proficients. Vanity, and a desire of broaching new Opinions, and rendering themselves remarkable, made everyone affect to have made new Discoveries in the Regions of the Air.

Vast Treatises were in a short time composed, and different Systems were every day set forth, some as distant from all Probability as they all were from one another. Every one had a particular Set of Followers, who appeared so well convinced of the Truth of what they professed that they declared themselves ready to endure Martyrdom for the Conviction of the rest. This puzzled the Minds of the People so much that they knew not to which to give Credit, and frequently occasioned Distinctions among them, to the ruin of many a noble Family.

For which reason, and also that the immoderate Application to Philosophy took our Youth from the more useful Studies of War, Politics, and Mechanism, Golbasto Momarin Eulame Guclo, the Father of our present Emperor, made an Edict, strictly prohibiting the use of Mathematics for the future, except in such Branches of it as were necessary for Navigation, or for Weight and Measure, with a Penalty of two thousand Gredulgribs (each of which in Gold is as big as a Silver Penny ˌofˌ English Money) affixed to the Conviction of the Crime after the Publication of this Order; and if the Delinquent was found incapable of paying such a Fine, his Life must answer for his Fault. All the Books of Argument relating to this Science were immediately burned, all the Paintings of it demolished, or plastered over, as you see it was here, and the same Punishment allotted for anyone who should conceal the one, or by any means preserve the other, as for him who should be guilty either by writing or painting a new one of the same kind. By this means, added he, Astronomy, Geography, and many other Branches of this noble Science, are entirely lost, or lie dormant in the Breasts of those who dare not transmit them to their Posterity.

Although no Man is a greater Admirer of these kind of Studies than myself, yet I confess I could not avoid thinking it very prudent in the Government of Lilliput to suppress them, when they began to encroach on the practical and more useful Business of Mankind while they live in the world. For, on mature Consideration, what is it to us by what means it first received its Formation, or how it is since influenced and directed, if we are provided with all things needful in it? And how void of Reason must we appear to a disinterested Observer, to lose that time in visionary Speculations, which is too little to be employed in the endeavour of acquiring what alone can defend us from Insults, and Contempt, and the want of those Necessaries, without which Life is so far from being desirable, that it becomes a Burden? In the midst of these more serious Reflections, I could not forbear laughing at a sudden Thought which just then came into my Head: that if such a Law were put in force in England, what a loss our Ladies would be at, for the Amusements they meet with in having their Fortunes told.

◦

ˌHeˌ again strictly charged me to blot out the Picture, as soon as I had enough considered it to be able to retain as much of it in my Mind as I thought would be of service either to the Amendment of my Morals, or Satisfaction of my Curiosity. I assured him, that I would obey him ˌ...ˌ.

Could I have taken it down, and brought it to England with me, I should have thought it a greater Treasure than all the Wealth of America; but that was impossible, not only on the account of the King's express Command ⌞...⌟ but also because it was painted on the Wall, which there was no removing without pulling down the Fabric, and by that means betraying the Theft.

Automathes

John Kirkby, 1745

An intriguing take on the desert island story, John Kirkby's novel depicts an individual whose shipwreck preserves (remarkably enough) little more than himself, a few books, and some mathematical instruments. He succeeds in teaching himself a range of skills including a good deal of mathematics, prompting reflections on the universality of mathematics and the possibility of devising an idiosyncratic form of it.

In one sense this book represents the converse of the various accounts we saw in Chapter 10 of the desirability of a mathematical education: for Automathes the acquisition of mathematical learning is simply unavoidable.

John Kirkby (c. 1705–1754), *The capacity and extent of the human understanding; exemplified in the extraordinary case of Automathes; a young nobleman, who was accidentally left in his infancy, upon a desolate island, and continued nineteen years in that solitary state, separate from all Human Society. A narrative Abounding with many surprizing Occurrences, both Useful and Entertaining to the Reader.* (London, 1745), pp. 165, 171–174, 177–187.

I was more fortunate in opening a Till in the End of the Chest, in which were deposited some Books and White Paper, a few Black-lead Pencils, Pen, and an Ink-Standish, a Pocket Magnifying-glass, a Seaman's Scale, and a Case of Mathematical Instruments ⌞...⌟

The Books were three in Number ⌞...⌟ *viz.* a Treatise of Divinity, a Piece of History, and a large System of the Mathematics. ⌞...H⌟ad it not been for the Cuts interspersed here-and-there, especially in the History, and the Geometrical Schemes, which abounded in the Mathematical Book, I believe I should have for ever laid them aside, without father Examination. But these, being more intelligible than the Letters, convinced me of their Design, and were afterwards a Direction for me to hold the Book. And the Mathematical Volume became of great Use to instruct me in the Principles of that Science, though without the least Knowledge of a Letter contained in it.

This taught me the Use of the Ruler, Compasses, and Brass Semicircle,° from their exact Resemblances upon the Paper with various ocular Applications of them; by which means I presently learned several manual Operations, such as to measure a ˌstraightˌ Line by equal Parts, to describe a circle, to bisect a ˌstraightˌ Line, erect and let fall a Perpendicular, to inscribe a regular Polygon of any proposed Number of Sides in a Circle, and to measure the Angles or Openings of two intersecting ˌstraightˌ Lines by the equal Divisions of an Arc.

I had also a frequent Sight of the nine Digits or Figures, both in the Book and upon the Instruments, and, observing the same Figures always to denote the same Numbers, I became at length perfectly acquainted with their Use, and likewise discovered how these alone, with the Character called a Cipher, were adapted to express all Numbers imaginable, by the Variation of their Places; A Contrivance, the exceeding Ingenuity of which, did not a little please me.

But though I had this Success with the Figures, yet it was impossible for me to have the like with the Letters of the Alphabet, which were put to signify Things I have not the least Notion of.

ˌ...ˌ I presently found out a Way, by Help of the Black-lead Pencils, to impress several Marks of my Skill in geometry upon the unwritten Paper, in which I made continual Improvements.

ˌ...ˌ I was discovered to have learned a Sort of Arithmetic peculiar to myself, and my Necessities had also taught me to perform several Problems in Geometry.

ˌ...ˌ Upon the highest Part of the Hill, directly behind my Cottage, a little Mount reared up its Head, in form of the lower Frustum of a Cone, overlooking the whole Country around, whose Summit made an horizontal Plain, of not more than ten Paces in Circumference. This I took the Pains to cover with a smooth Cap of Clay, erecting a Stile near the Centre of about a Yard in Height, which was my Contrivance to trace out the Points of Shadow, according to the different Times of the Day, and Seasons of the Year, that I might attain a more perfect Knowledge of the Motions of the Sun. But here I had many Experiments before I could answer my Design, till at last I found out a way to make my Stile out of a Piece of the Copper Skillet, drilling an Hole through the Top to admit the Rays of the Sun; by which Invention I both obtained a more certain Perpendicular to my Plane, and a truer Method of marking the Points of Shadow from a lucid Point, without any perceptible Penumbra.

This daily Practice of marking out the Points of Shadow, presently suggested to me a Meridian Line from the Foot of the Stile, in which I saw every Day's shortest (or Mid-day) Shadow constantly to fall. And I observed that all Lines drawn from the same Centre, making equal Angles on both Sides ₍of₎ the Meridian, must every Day coincide with the Shadow of the Stile, the same Spaces of Time before and after Noon, and that the Length of the Mid-day Shadow alone was sufficient to mark out the annual Access and Regress of the Sun. I made me therefore a Circle from the Foot of the Stile as a Centre, to as great a Circumference as my Plane would admit, the Northern Limb of which I divided into five equal Arcs, of twenty Degrees apiece, on both Sides ₍of₎ the Meridian, with Lines drawn from the Centre to each; and thus I had the Angles of Shadow marked out for an Hundred Degrees, on each Side ₍of₎ the Meridian. But I presently perceived the Insufficiency of this to measure Time, which was the End I aimed at by it, from the greater or lesser Disproportion between these Angles, and the like Arcs of the Sun's course, according as they were more or less distant from the Meridian. And this I had no Remedy for, till after much Study I invented a Time-teller out of one of my Boxes.

I chose out the compacter of these, which, after I had emptied of what it contained into the Chest, I endeavoured to make as tight as possible to contain Water; and, when I had brought it to my Mind, I made a Hole in the Bottom, fitting it with a Stopple. Afterwards I proceeded, with all the Care and Exactness I was Master of, to make another horizontal Plane, close by the Fountain, drawing a Circle, and erecting a perpendicular Stile in the Center, both of the same Dimensions with that on the Mount. And having found the Meridian Line, I placed my Box in a right horizontal Position across the Outlet of the Fountain, filling it with Water. Then waiting till the Sun came upon the Meridian, I immediately let out the Water, and, as soon as the Box was empty, marked down the Shadow upon the Circle, filling it again out of the Fountain, and placing it in the same Situation. All which was done with such Expedition, that the Box was scarce empty before it was full again, and the Orifice was kept continually running. This I kept repeating till the Setting of the Sun, still marking down the Points of Shadow, at the End of every Box-full thus run out, and laying the same

Extent with my Compasses upon the Circle to the Eastward, which the Shadow made to the Westward. And by that means I obtained a pretty exact Dial within my Enclosure, for almost Ninety Degrees on both sides of the Meridian, or from near six in the Morning to six in the Evening, which I afterwards increased at my Leisure, as the Days grew longer.

When I had brought this to as great a Perfection as I was able, I expunged all my Lines of Shadow, save the Meridian, upon the Mount, and laid down the same with these in their Places, having there a more commodious Situation for my intended Observations of the Heavenly Bodies. Out of the Lid of the Water-box I formed a Quadrant, of about two Foot Radius, graduated as I had seen in the Book, which, upon Occasion, I applied perpendicularly along the Meridian (with the Limb towards the Stile, and the Centre upon the Extremity of the lucid Ray) to measure the different Altitudes of the Sun in its Mid-day Situation. The same also served me to take the Meridian Altitudes of the Moon, which I performed by leaning along the Side of the Mount, where I made a Place to fix my Feet, and drawing the Centre to my Eye fixed against the North Edge of the Plane. And I was so intent upon these two great Luminaries, that I seldom carried my Observations of this Sort to the Stars, perceiving none of them, except four or five, but what always kept the same Situation one to another, so that whenever I obtained the true Altitude of one, I could refer all the rest to it in their Order.

By this frequent Practice I learned the gradual Increase and Decrease of Day and Night, according to the Access and Regress of the Sun, as also the Number, nearly, of diurnal Revolutions, which he took up in finishing his annual Course. I determined the menstrual Periods of the Moon, and from the exact Conformity I beheld between the Changes of her Face, and her Distances from the Sun, I could, at any time, compute her Age, by a Circle contrived for that Purpose. And from the frequent Experience of Eclipses, both of the Sun and Moon, I also arrived to a true Notion of their Causes, being perfectly instructed in the borrowed Light of the latter from the former. So far proceeded my Astronomy.

And these constant Exercises in the Description of Lines, Circles, and Angles, brought me likewise to an Acquaintance with several Truths in speculative Geometry; for which Purpose, after my Paper was spent, I expunged the Lines out of my Dial by the Fountain, and drew all my Schemes upon that, inserting and blotting out at Pleasure. I perceived the

Proportionality of the ⌞corresponding⌟ Sides of similar Figures, and the Equality between all the alternate (as well as between all the opposite) Angles made by the Intersection of the same ⌞straight⌟ Line with any Number of Parallels. I learned the Equality of the Angles in a Triangle to a Semicircle, and of any outward Angle in the same to the two inward opposite ones. I found out the ⌞1 : 2⌟ Proportion which any Angle, at the Circumference of a Circle, bears to one at the Centre, standing upon the same Base. I discovered the Equality of the Areas of all Parallelograms upon the same Base between the same Parallels, which put me also upon studying the Proportions between other dissimilar Surfaces; and I likewise attained to the Knowledge of Pythagoras's Theorem of the Equality between the Square of the longest Side in a right-angled Triangle, and the Sum of the Squares of the two shortest.

And as Lines led me to consider the Surfaces of which they were the Limits, so did these Surfaces bring me to an Acquaintance with the Solids bounded by them, for the right Understanding of which, I found great Helps from the actual Formation of several regular Bodies out of Clay, though the Draughts themselves, which I found of these in the Book, were adapted in the best Manner to assist the Imagination, as being delineated according to the exactest Rules of Perspective, and everywhere skilfully shaded. I observed the ⌞squared ratio⌟, which similar Surfaces, and the ⌞cubed ratio⌟, which similar Solids, had ⌞with respect to the ratio of⌟ their Altitudes, or ⌞corresponding⌟ Sides; and as I naturally pitched upon the Square for the Mensuration of the former, so did I apply the Cube for the Standard of these latter. I learned the chief Properties of the Sphere, the Cone, and the Cylinder, and became intimately conversant with the five regular Bodies. I formed an Ellipse from the oblique Section of a Cylinder, and had some Notion of the other Apollonian Curves,° from the actual Sections of a Cone.

I also made some Progress in the more abstract Theory of Numbers, as well as in the more sensible Speculation of Magnitude. I was necessarily led to that prime Distinction of Number, into Integer and ⌞fraction⌟, according as the Quantities they were brought to express were discontinu⌞ous⌟, or continu⌞ous⌟, and I considered them likewise in respect of their Composition as Even and Odd, Prime and Composite, and compared them also together, as Commensura⌞ble⌟ and Incommensura⌞ble⌟. I deduced the principal Truths which belong to the Doctrine of Proportions, and arrived to some Skill in Progressions, both Arithmetical and Geometrical. But yet I would have you take notice that I attribute all those high Acquirements in Numbers, as well as Magnitudes, rather to the Assistance of the Book, than my own Invention alone. For when I had attained to the Knowledge of the numeral Figures, as you have already heard, I seldom failed, through the whole Work, to fish out the Author's intentions by them, though I understood not the Meaning of a Word he wrote.

Notes

Brass Semicircle: Kirkby probably means something like a modern protractor for measuring angles.

Apollonian Curves: the conic sections; that is, the ellipse, parabola, and hyperbola produced by slicing a cone at various angles.

The Loves of the Triangles

John Frere, 1798

John Frere's spoof of Erasmus Darwin's botanical poem *The Loves of the Plants* sheds a pleasing sidelight on the popular image of mathematics at the end of the eighteenth century; like Cavendish and the anonymous writer in *Fun*, Frere picks up on the incomprehensibility of mathematical jargon to outsiders. Frere was a diplomat and a poet and translator.

John Hookham Frere (1769–1846), *The Loves of the Triangles. A Mathematical and Philosophical Poem.* In *The anti-Jacobin*, 1798; republished in *The Works* (London, 1872), vol. I, pp. 88–93.

Stay your rude steps, or e'er your feet invade
The Muses' haunts, ye sons of War and Trade!
Nor you, ye legion fiends of Church and Law,
Pollute these pages with unhallow'd paw!
Debased, corrupted, grovelling, and confined,
No Definitions touch your senseless mind;
To you no Postulates prefer their claim,
No ardent Axioms your dull souls inflame;
For you, no Tangents touch, no Angles meet,

No Circles join in osculation sweet!
For me, ye Cissoids, round my temples bend
Your wandering curves; ye Conchoids extend;
Let playful Pendules quick vibration feel,
While silent Cyclois rests upon her wheel;
Let Hydrostatics, simpering as they go,
Lead the light Naiads on fantastic toe;
Let shrill Acoustics tune the tiny lyre;
With Euclid sage fair Algebra conspire;
The obedient pulley strong Mechanics ply,
And wanton Optics roll the melting eye!
I see the fair fantastic forms appear,
The flaunting drapery, and the languid leer;
Fair sylphish forms—who, tall, erect, and slim,
Dart the keen glance, and stretch the length of limb;
To viewless harpings weave the meanless dance,
Wave the gay wreath, and titter as they prance.

Such rich confusion charms the ravish'd sight,
When vernal Sabbaths to the Park invite.
Mounts the thick dust, the coaches crowd along,
Presses round Grosvenor Gate th'impatient throng;
White-muslined misses and mammas are seen,
Link'd with gay cockneys, glittering o'er the green:
The rising breeze unnumber'd charms displays,
And the right ankle strikes th' astonished gaze.

But chief, thou Nurse of the Didactic Muse,
Divine Nonsensia, all thy soul infuse;
The charms of Secants and of Tangents tell,
How Loves and Graces in an Angle dwell;
How slow progressive Points protract the Line,
As pendent spiders spin the filmy twine;
How lengthen'd Lines, impetuous sweeping round,
Spread the wide Plane, and mark its circling bound;
How Planes, their substance with their motion grown,
Form the huge Cube, the Cylinder, the Cone.

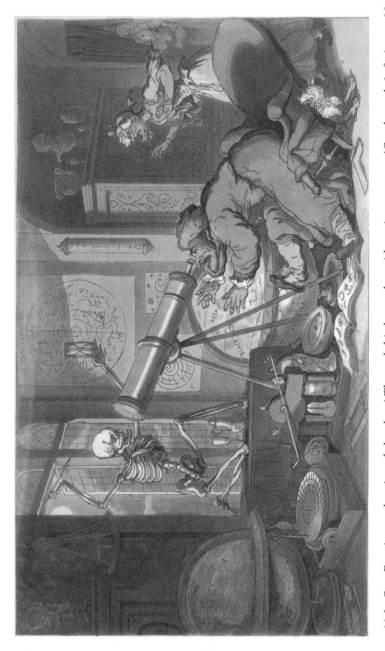

Figure 11.2. One Evening, as he viewed the sky, / Through his best tube, with curious eye . . . . (Combe, vol. 2, facing p. 38. © Princeton University Library. Rare Books Division, Department of Rare Books and Special Collections, 3686.7.332.)

Master Senex the Astronomer

William Combe, 1815

William Combe is best known for the "Doctor Syntax" series of comic poems, an illustrated satire upon ideas of the "picturesque" which came toward the end of his long career as a writer, principally of satire and history. Syntax's quixotic misadventures filled four installments, of which *The English Dance of Death*, from which this extract is taken, was the second. The writing is basically light-hearted, but "the astronomer" does not come across at all well.

William Combe (1742–1823), "The Astronomer" in *The English Dance of Death* (London, 1815), vol. 2, pp. 38–42.

He, who with care and much ado,
Has changed one blade of grass to two;
He, who an acre too has ploughed,
And with good seed that acre sowed;
He, who to the Earth has given
A Tree, to rear its boughs to Heaven,
And, with a chaste and loving wife,
Gives but a single babe to life,
Has, as 'tis said, by one whose name
Stands foremost on the roll of Fame,
Performed, in philosophic view,
All that a Man's required to do.
This done, each social claim is paid,
And when in Earth his bones are laid,
The sculptured stone may truly tell
That he has lived and acted well.

But what says Science to the Rule
Thus taught in simple Nature's school:
That Science which pursues her way,
Through gloomy night, or glaring day,
Creation's every work explores,

Digs deep for all the hidden stores
Which the Earth's darksome caves contain,
And dives within the watery main,
Expatiates through the fields of air,
And sees the storms engendered there,
Or boldly bids her daring eye
Explore the wonders of the sky;
While Genius, to no spot confined,
That brightest offspring of the mind,
Ranges at will, through Space and Time,
In every age, in every clime,
And oft, its glorious toil to crown,
Creates new Systems of its own.
—Such are the classes that embrace
Man's social, cultivated Race;
And, as each acts the part assigned,
It helps, in due degree, to bind,
By harmonising, just control,
The general order of the whole.

Now Master Senex, who was bred
To guide into the youthful head,
Not that poor Two and Two make Four,
Or that three Twenties form Threescore,
But the nice, calculating play
Of Decimals and Algebra,
With Problems and the curious store
That's found in Mathematic Lore,
He always felt himself at home
When 'mong the Stars he chose to roam,
And, for a frisk, would sometimes stray
Delighted in the Milky Way.
Would bask in the Meridian Noon,
And clamber Mountains in the Moon.
He would the Comet's course pursue,
And tell, with calculation due,
How many million miles it posted,

While a small Leg of Mutton roasted,
And how many a thousand years
Will pass before it re-appears.

—He never for one moment thought
But of the Sciences he taught;
Him never did the Fancy seize
Of ploughing land, or planting trees;
Nor was the sober Sage beguil'd
To be the Father of a Child.
A Sister, an old saving Elf,
Who was as barren as himself,
Added a figure to the scene,
And dressed his meat, and kept him clean.
One Evening, as he viewed the sky
Through his best tube, with curious eye,
And 'mid the azure wilds of air
Pursued the progress of a Star,
A Figure seemed to intervene,
Which in the sky he ne'er had seen,
But thought it some new planet given,
To dignify his views of Heaven.
"O this will be a precious boon!
Herschel's Volcanoes in the Moon,°
Are nought to this," Old Senex said,
"My Fortune is for ever made."
—"It is, indeed," a voice replied;
The Old Man heard it—terrified,
And, as Fear threw him to the ground,
Through the long tube Death gave the wound.

Though Senex died, no thunder rolled,
No lightning flashed, no tempests growled;
Nor did the Pleiades descend,
In rain, to weep their faithful friend;
Nor would the Moon in sorrow shroud
Her silver light within a cloud;

Nay, not a single sigh was given
By any Star that shines in Heaven.

Note

Herschel's Volcanoes in the Moon: William Herschel (1738–1822), the famous astronomer,
published a paper on the mountains on the moon in the *Philosophical Transactions of the
Royal Society* in 1780.

An Ode to the Mathematics

Alfred Domett, 1833

Alfred Domett had a varied career as a New Zealand politician and administrator,
including a period as premier in 1862–1863. In youth and retirement he published volumes of verse in England; this poem is dated in 1830, when he was
nineteen. Seemingly directed solely at the author's own mathematical incapacity,
it touches the theme of the incomprehensibility of mathematics which we have
met elsewhere, without reaching the harshness of Combe or even Cavendish in
condemning the subject itself.

Alfred Domett (1811–1887), "Ode to the Mathematics: A Cambridge Ebullition," in *Poems*
(London, 1833), pp. 52–53.

> Ye Mathematics! over which I pore
>   Full stolidly—yet to my sorrow find
> I cannot fix upon your crabbèd lore
>   My scape-grace, wandering, weak, wool-gathering mind—
> Oh, are yet not in language plain, a bore,
>   For luckless wight like me a plague refined—
> Ye intellectual catacombs—where drones
> Of many an age have piled up musty bones!
>
> We are old foes—yet can't, it seems, be loosed
>   From one another, though we tug the chain
> Like coupled hounds; I have so oft abused
>   And railed at you, and yet returned again

To be by your dark mysteries confused,
  That at my fate I smile, and friendship fain
Would offer—foes are well-nigh friendly (trust 'em),
Whose regular abuse becomes a custom.

They say you lead to grand results, and Science
  Makes you unto her heaven a Jacob's ladder;
So clouded though, we cannot see the sky hence,
  And black-gowned students are a vision sadder,
Nor promise half so much for what they spy hence,
  As did the white-robed angels Jacob had a
Glimpse of; but be that as it may, you lead to
Things greater far than I can e'er give heed to.

You teach the stars—their courses—like Silenus,°
  Teach what the world is set a-going by,
And all the eccentricities of Venus;
  You compass earth and ocean, land and sky,
Teach us to argue and to squabble, wean us
  From base delights (no doubt) to pure and high;
You teach mankind all, all that can ennoble 'em;
Meantime I'm staggered by this plaguy problem!

Note

Silenus: the mythological companion to the wine god Dionysos variously appeared in myths
  as tutor to Dionysos, adviser to King Midas, and a commentator on earthly rulers.

"Some Veritable Urania"

Augusta Jane Evans, 1864

Augusta Jane Evans, popular novelist of the American South, produced some
magnificent depictions of women intellectuals. Here, in what is surely one of the
best such passages, she describes one heroine's devotion to astronomy.

Augusta J. Evans (1835–1909), *Macaria: or, Altars of Sacrifice* (Richmond: 2nd edition, 1864),
pp. 77–79.

"Irene, it is past midnight."

She gave no intimation of having heard him.

"Irene, my child, it is one o' clock."

Without looking up she raised her hand toward the clock on the mantle, and answered, coldly:

"You need not sit up to tell me the time of night; I have a clock here. Go to sleep, Uncle Eric."

He rested his shoulder against the door-facing, and, leaning on his crutches, watched her.

She sat there just as he had seen her several times before, with her arms crossed on the table, the large celestial globe drawn near, astronomical catalogues scattered about, and a thick folio open before her. She wore a loose wrapper, or robe de chambre, of black velvet, lined with crimson silk and girded with a heavy cord and tassel. The sleeves were very full, and fell away from the arms, exposing them from the dimpled elbows, and rendering their pearly whiteness more apparent by contrast with the sable hue of the velvet, while the broad round collar was pressed smoothly down, revealing the polished turn of the throat. The ivory comb lay on the table, and the unbound hair, falling around her shoulders, swept over the back of her chair and trailed on the carpet. A miracle of statuesque beauty was his queenly niece, yet he could not look at her without a vague feeling of awe, of painful apprehension; and, as he stood watching her motionless figure in its grand yet graceful pose, he sighed involuntarily. She rose, shook back her magnificent hair, and approached him. Her eyes, so like deep, calm azure lakes, crossed by no ripple, met his, and the clear, pure voice echoed through the still room.

"Uncle Eric, I wish you would not sit up on my account; I do not like to be watched."

"Irene, your father forbade your studying until this hour. You will accomplish nothing but the ruin of your health."

"How do you know that? Do statistics prove astronomers short-lived? Rather the contrary. I commend you to the contemplation of their longevity. Good-night, Uncle; starry dreams to you."

"Stay, child; what object have you in view in all this laborious investigation?"

"Are you sceptical of the possibility of a devotion to science merely for science's sake? Do my womanly garments shut me out of the Holy of

Holies, debar me eternally from sacred arcana, think you? Uncle Eric, once for all, it is not my aim to

'—brush with extreme flounce
The circle of the sciences.'

"I take my heart, my intellect, my life, and offer all upon the altar of its penetralia. You men doubt women's credentials for work like mine; but this intellectual bigotry and monopoly already trembles before the weight of stern and positive results which women lay before you—data for your speculations—alms for your calculation. In glorious attestation of the truth of female capacity to grapple with some of the most recondite problems of science stand the names of Caroline Herschel, Mary Sommerville, Maria Mitchel, Emma Willard, Mrs. Phelps, and the proud compliment paid to Madame Lepaute° by Clairaut and Lalande, who, at the successful conclusion of their gigantic computations, declared: 'The assistance rendered by her was such that, without her, we never should have dared to undertake the enormous labor in which it was necessary to calculate the distance of each of the two planets, Jupiter and Saturn, from the comet, separately for every degree, for one hundred and fifty years.' Uncle Eric, remember

" '—Whoso cures the plague,
Though twice a woman, shall be called a leech;
Who rights a land's finances is excused
For touching coppers, though her hands be white.' "

She took the volume she had been reading, selected several catalogues from the mass, and, lighting a small lamp, passed her uncle and mounted the spiral staircase leading to the observatory. He watched her tall form slowly ascending, and, in the flashing light of the lamp she carried, her black dress and floating hair seemed to belong to some veritable Urania— some ancient Egyptic Berenice.° He heard her open the glass door of the observatory, then the flame vanished, and the click of the lock fell down the dark stairway as she turned the key. With a heavy sigh the cripple returned to his room, there to ponder the singular character of the woman whom he had just left, and to dream that he saw her transplanted to the constellations, her blue eyes brightening into stars, her waving hair braiding itself out into brilliant, rushing comets. The night was keen, still, and cloudless, and, as Irene locked herself in, the chill from the marble tiles crept through the carpet to her slippered feet. In the centre of the apartment rose a wooden shaft bearing a brass plate, and to this a telescope was securely fastened. Two chairs and an old-fashioned oaken table, with curious carved legs, comprised the furniture. She looked at the small siderial clock, and finding that a quarter of an hour must elapse before she could make the desired observation, drew a chair to the table and seated herself. She took from the drawer a number of loose papers, and prepared the blank-book for registering the observation, then laid before her a slate covered with figures, and began to run over the calculation. At the close of fifteen minutes she placed herself at the telescope, and waited patiently for the appearance of a small star which gradually entered the field; she noted the exact moment and position, transferred the result to the register, and after a time went back to slate and figures. Cautiously she went over the work, now and then having recourse to pen and paper; she reached the bottom of the slate and turned it over, moving one finger along the lines. The solution was wrong; a mistake had been made somewhere; she pressed her palm on her forehead, and thought over the whole question, then began again. The work was tedious, the calculation subtle, and she attached great importance to the result; the second examination was fruitless as the first; time was wearing away; where could the error be? Without hesitation she turned back for the third time, and commenced at the first, slowly, patiently threading the maze. Suddenly she paused and smiled; there was the mistake, glaring enough, now. She corrected it, and working the sum through, found the result perfectly accurate, according fully with the tables of Leverrier by which she was computing. She carefully transferred the operation from slate to paper, and, after numbering the problem with great particularity, placed all in the drawer and turned the key. It was three o' clock; she opened the door, drew her chair out on the little gallery, and sat down, looking toward the East. The air was crisp but still, unswayed by current waifs; no sound swept its crystal waves save the low, monotonous distant thunder of the Falls, and the deep, cloudless blue ocean of space glowed with its numberless argosies of stellar worlds. Constellations which, in the purple twilight, stood sentinel at the horizon, had marched in majesty to mid-heaven, taken reconnoissance thence, and as solemnly passed the opposite horizon to report to watching gazers in another hemisphere. "Scouts stood upon every headland, on every plain"; mercilessly the inquisitorial eye of science followed the heavenly wanderers; there was no escape from the eager, sleepless police who kept vigil in every clime and country; as well call on Böotes to give o'er his care of Ursa Major, as hopelessly attempt to thrust him from the ken of Cynosura.° From her earliest recollection, and especially from the hour of entering school, astronomy and mathematics had exerted an overmastering influence upon Irene's mind. The ordinary text-books only increased her interest in the former science, and while in New York, with the aid of the professor of astronomy, she had possessed herself of all the most eminent works bearing upon the subject, sending across the Atlantic for tables and selenographic charts which were not to be procured in America.

Under singularly favorable auspices she had pursued her studies perseveringly, methodically, and, despite her father's prohibition, indefatigably. He had indulged, in earlier years, a penchant for the same science, and cheerfully facilitated her progress by rearranging the observatory so as to allow full play for her fine telescope; but, though proud of her proficiency, he objected most strenuously to her devoting so large a share of her time and attention to this study, and had positively interdicted all observations after twelve o' clock. Most girls patronize certain branches of investigation with fitful, spasmodic vehemence, or periodic impulses of enthusiasm; but Irene knew no intermission of interest, she hurried over no details, and, when the weather permitted, never failed to make her nightly visit to the observatory. She loved her work as a painter his canvas, or the sculptor the marble one day to enshrine his cherished ideal; and she prosecuted it, not as a mere pastime, not as a toy, but as a life-long labor, for the labor's sake.

Notes

Caroline Herschel, Mary Sommerville, Maria Mitchel, Mrs. Phelps, and Madame Lepaute: Women scientists and astronomers of the late eighteenth and nineteenth centuries. Emma Willard was a pioneer of women's higher education.

Urania: the muse of astronomy. Berenice II, queen of Egypt in the third century BC, was the subject of a famous story in which her hair, cut off and placed in the temple of Aphrodite, disappeared and was said to have been placed among the stars. The constellation coma berenices still appears on star maps.

Böotes and Ursa Major are (adjacent) constellations, traditionally the heavenly transformations of Arcas and his mother Callisto; Cynosura is the pole star.

Fun

1863, 1870

Founded in 1861 to compete with *Punch*, the London magazine *Fun* appeared weekly with a mix of satire, parody, and political, literary, and other reportage until 1901; its best-known contributor was W. S. Gilbert. The second of these passages is a response to J. J. Sylvester's *Laws of Verse*, mentioned in Chapter 10.

Fun, Saturday, March 28, 1863, p. 19; Saturday, September 10, 1870, p. 98.

Mathematical Problems

Required—

To sit comfortably in a chair, the back and seat of which are at right angles.

To know when skating whether a plane ⌐surface⌐ hath length and breadth by personal experience.

To find the centre of a family circle when there is cold meat for dinner, and a few female friends of some benevolent society coming to tea in the evening.

To find the centre of gravity when Sothern° is performing at the Haymarket.

To feel gratified on seeing in the mirror over the drawing-room mantel-piece after (as you thought) a telling *entré*, the segment of a circle beautifully imprinted on your brow from the rim of your hat.

To find that sticking the fish-hook into the fleshy part of your arm is anything but an acute angle.

To shave with your razor at a right angle without cutting your throat, or that, if you do so partially, you think it a right angle.

Mathematical Poetry

To the Editor of *Fun*.

Sir, I see that Professor Sylvester, the famous mathematician, has written a book, entitled *The Laws of Verse*, to prove that poetry is as exact a science as mathematics, and that any fellow who has passed the *Pons Asinorum*°

can write an epic, and that after all an algebraic equation is very much the same thing as a sonnet. Hurrah! will be the cry of our lads, who will naturally prefer taking their Euclid in the form of *Don Juan*, while the children will promote "Dickery, dickery dock" over the multiplication table.

But no matter! I have turned my attention to the subject, and although I have not seen the learned professor's work, I think I have hit on the method. Here is my first poem:—

$$a + a^2 \times 3\sqrt{m - n} = \frac{1}{l} + x \left(a + b - \frac{c}{d} \right).$$

Of course this equation has to be worked out, but I will not occupy your valuable space by so doing, as it is a thing which any schoolboy—especially if he knows mathematics remarkably well—will effect with ease.

Or, to put it geometrically instead of algebraically, let ABC be a given line. From the centre B at the distance DEF describe a circle, whose square shall be equal to the squares of AB, BC together with twice the rectangle contained by BE, DF.

Anyhow, here is the poem:

I never knew a wild kodoo
    Analysis, Zitella,
Tobacco-box, or kangaroo,
    Or infantile umbrella
        Could dance a jig
        Like the learned pig,
    Or even a Tarantella!

For why, I have never visited
    The coasts of Willy-Nilly,
Or the cataract-head of Slugabed,
    Or the billowy shores of Chili.
        Nor sailed a spoon
        To the crescent moon
    With a cargo of picalilly!

So I tear my hair in blind despair,
    And blister my eyes with peaches,

Fling sea-anemones everywhere
    In anti-macassar trousers,°
        And my soul I vex
        With "$n + x$,"
    And mathematical speeches.

    This, if not according to Professor Sylvester, at least according to Euclid is mathematical verse, for it is composed of "lines, having length without breadth."
Yours,
A Ninny to the Power of Nine.

Notes

Sothern: Edward Askew Sothern (1826–1881), comic actor. The Haymarket was (and is) a major London theatre.

Pons Asinorum: Proposition 5 of Book 1 of Euclid's *Elements*: the angles opposite the equal sides of an isosceles triangle are equal.

In anti-macassar trousers: "There's some mistake here, owing to the difficulty of ascertaining the value of the symbol $\frac{man}{9}$" (footnote appearing in the original).

A Sight of Thine Interior

Edwin A. Abbott, 1884

Edwin Abbott Abbott, Anglican priest, teacher, and writer, wrote works of theology, grammar, and history and edited the *Essays* of Francis Bacon, as well as serving for more than two decades as the inspiring headmaster of the City of London School. But he is best known as "A. Square," the author of the science fiction classic *Flatland*, which recounts the adventures of an inhabitant of a two-dimensional world given glimpses of both higher and lower dimensions. The book combined sharply pointed satire and Christian teaching with serious mathematics and has inspired a multitude of other works: adaptations, sequels, and films.

    In this passage, A. Square, intoxicated by his view of a three-dimensional world, begins to speculate about worlds with still more dimensions. His audacity does not go down very well, and in his own world it will eventually land him in prison.

A. Square [Edwin A. Abbott (1838–1826)], *Flatland: A Romance of Many Dimensions* (London, 1884), pp. 85–89.

The Sphere would willingly have continued his lessons by indoctrinating me in the conformation of all regular Solids, Cylinders, Cones, Pyramids, Pentahedrons, Hexahedrons, Dodecahedrons and Spheres: but I ventured to interrupt him. Not that I was wearied of knowledge. On the contrary, I thirsted for yet deeper and fuller draughts than he was offering to me.

"Pardon me," said I, "O Thou Whom I must no longer address as the Perfection of all Beauty; but let me beg thee to vouchsafe thy servant a sight of thine interior."

SPHERE. "My what?"

I. "Thine interior: thy stomach, thy intestines."

SPHERE. "Whence this ill-timed impertinent request? And what mean you by saying that I am no longer the Perfection of all Beauty?"

I. My Lord, your own wisdom has taught me to aspire to One even more great, more beautiful, and more closely approximate to Perfection than yourself. As you yourself, superior to all Flatland forms, combine many Circles in One, so doubtless there is One above you who combines many Spheres in One Supreme Existence, surpassing even the Solids of Spaceland. And even as we, who are now in Space, look down on Flatland and see the insides of all things, so of a certainty there is yet above us some higher, purer region, whither thou dost surely purpose to lead me—O Thou Whom I shall always call, everywhere and in all Dimensions, my Priest, Philosopher, and Friend—some yet more spacious Space, some more dimensionable Dimensionality, from the vantage-ground of which we shall look down together upon the revealed insides of Solid things, and where thine own intestines, and those of thy kindred Spheres, will lie exposed to the view of the poor wandering exile from Flatland, to whom so much has already been vouchsafed.

SPHERE. Pooh! Stuff! Enough of this trifling! The time is short, and much remains to be done before you are fit to proclaim the Gospel of Three Dimensions to your blind benighted countrymen in Flatland.

I. Nay, gracious Teacher, deny me not what I know it is in thy power to perform. Grant me but one glimpse of thine interior, and I am satisfied for ever, remaining henceforth thy docile pupil, thy unemancipable slave, ready to receive all thy teachings and to feed upon the words that fall from thy lips.

SPHERE. Well, then, to content and silence you, let me say at once, I would show you what you wish if I could; but I cannot. Would you have me turn my stomach inside out to oblige you?

I. But my Lord has shown me the intestines of all my countrymen in the Land of Two Dimensions by taking me with him into the Land of Three. What therefore more easy than now to take his servant on a second journey into the blessed region of the Fourth Dimension, where I shall look down with him once more upon this land of Three Dimensions, and see the inside of every three-dimensioned house, the secrets of the solid earth, the treasures of the mines in Spaceland, and the intestines of every solid living creature, even of the noble and adorable Spheres.

SPHERE. But where is this land of Four Dimensions?

I. I know not: but doubtless my Teacher knows.

SPHERE. Not I. There is no such land. The very idea of it is utterly inconceivable.

I. Your Lordship tempts his servant to see whether he remembers the revelations imparted to him. Trifle not with me, my Lord; I crave, I thirst, for more knowledge. Doubtless we cannot *see* that other higher Spaceland now, because we have no eye in our stomachs. But, just as there was the realm of Flatland, though that poor puny Lineland Monarch could neither turn to left nor right to discern it, and just as there was close at hand, and touching my frame, the land of Three Dimensions, though I, blind senseless wretch, had no power to touch it, no eye in my interior to discern it, so of a surety there is a Fourth Dimension, which my Lord perceives with the inner eye of thought. And that it must exist my Lord himself has taught me. Or can he have forgotten what he himself imparted to his servant?

In One Dimension, did not a moving Point produce a Line with *two* terminal points?

In Two Dimensions, did not a moving Line produce a Square with *four* terminal points?

In Three Dimensions, did not a moving Square produce—did not this eye of mine behold it—that blessed Being, a Cube, with *eight* terminal points?

And in Four Dimensions shall not a moving Cube—alas, for Analogy, and alas for the Progress of Truth, if it be not so—shall not, I say, the motion of a divine Cube result in a still more divine Organization with *sixteen* terminal points?

Behold the infallible confirmation of the Series, 2, 4, 8, 16: is not this a Geometrical Progression? Is not this—if I might quote my Lord's own words—"strictly according to Analogy"?

Again, was I not taught by my Lord that as in a Line there are *two* bounding Points, and in a Square there are *four* bounding Lines, so in a Cube there must be *six* bounding Squares? Behold once more the confirming Series, 2, 4, 6: is not this an Arithmetical Progression? And consequently does it not of necessity follow that the more divine offspring of the divine Cube in the Land of Four Dimensions, must have 8 bounding Cubes: and is not this also, as my Lord has taught me to believe, "strictly according to Analogy"?

O, my Lord, my Lord, behold, I cast myself in faith upon conjecture, not knowing the facts; and I appeal to your Lordship to confirm or deny my logical anticipations. If I am wrong, I yield, and will no longer demand a Fourth Dimension; but, if I am right, my Lord will listen to reason.

I ask therefore, is it, or is it not, the fact, that ere now your countrymen also have witnessed the descent of Beings of a higher order than their own, entering closed rooms, even as your Lordship entered mine, without the opening of doors or windows, and appearing and vanishing at will? On the reply to this question I am ready to stake everything. Deny it, and I am henceforth silent. Only vouchsafe an answer.

SPHERE (AFTER A PAUSE). It is reported so. But men are divided in opinion as to the facts. And even granting the facts, they explain them in different ways. And in any case, however great may be the number of different explanations, no one has adopted or suggested the theory of a Fourth Dimension. Therefore, pray have done with this trifling, and let us return to business.

I. I was certain of it. I was certain that my anticipations would be fulfilled. And now have patience with me and answer me yet one more question, best of Teachers! Those who have thus appeared—no one knows whence—and have returned—no one knows whither—

have they also contracted their sections and vanished somehow into that more Spacious Space, whither I now entreat you to conduct me?

SPHERE (MOODILY). They have vanished, certainly—if they ever appeared. But most people say that these visions arose from the thought—you will not understand me—from the brain; from the perturbed angularity of the Seer.

I. Say they so? Oh, believe them not. Or if it indeed be so, that this other Space is really Thoughtland, then take me to that blessed Region where I in Thought shall see the insides of all solid things. There, before my ravished eye, a Cube, moving in some altogether new direction, but strictly according to Analogy, so as to make every particle of his interior pass through a new kind of Space with a wake of its own—shall create a still more perfect perfection than himself, with sixteen terminal Extra-solid angles, and Eight solid Cubes for his Perimeter. And once there, shall we stay our upward course? In that blessed region of Four Dimensions, shall we linger on the threshold of the Fifth, and not enter therein? Ah, no! Let us rather resolve that our ambition shall soar with our corporal ascent. Then, yielding to our intellectual onset, the gates of the Sixth Dimension shall fly open; after that a Seventh, and then an Eighth—

How long I should have continued I know not. In vain did the Sphere, in his voice of thunder, reiterate his commands of silence, and threaten me with the direst penalties if I persisted. Nothing could stem the flood of my ecstatic aspirations. Perhaps I was to blame; but indeed I was intoxicated with the recent draughts of Truth to which he himself had introduced me. However, the end was not long in coming. My words were cut short by a crash outside, and a simultaneous crash inside me, which impelled me through Space with a velocity that precluded speech. Down! down! down! I was rapidly descending; and I knew that return to Flatland was my doom. One glimpse, one last and never-to-be-forgotten glimpse I had of that dull level wilderness—which was now to become my Universe again—spread out before my eye. Then a darkness. Then a final, all-consummating thunder-peal; and, when I came to myself, I was once more a common creeping Square, in my Study at home, listening to the Peace-Cry of my approaching Wife.

Scenes in the Life of Pythagoras

Geoffrey Willans and Ronald Searle, 1953

A development of the twentieth century was the light-hearted take on the "greats" of the mathematical past, and few were more irreverent than that of the remarkable schoolboy Nigel Molesworth, the superb comic creation of Geoffrey Willans and Ronald Searle. Willans was a modestly successful author of (mostly) humorous nonfiction, Searle one of the great cartoonists and illustrators of the twentieth century, his work ranging from harrowing depictions of conditions in World War II prisoner-of-war camps to the well-known belles of St. Trinian's. Figure 11.3 shows his take on classical geometry.

The spelling is that of the original.

Geoffrey Willans (1911–1958) and Ronald Searle (b. 1920), *Molesworth* (London, 1992; originally published 1953), pp. 45–46.

Pythagoras as a mater of fact is at the root of all geom. Instead of growing grapes figs dates and other produce of greece Pythagoras aplied himself to triangles and learned some astounding things about them which hav been inflicted on boys ever since.

Whenever he found a new thing about a triangle Pythagoras who had no shame jumped out of his bath and shouted "Q.E.D." through the streets of athens its a wonder they never locked him up.

To do geom you hav to make a lot of things equal to each other when you can see perfectly well that they don't. This agane is due to Pythagoras and it formed much of his conversation at brekfast.

PYTHAGORAS (HELPING HIMSELF TO PORRIDGE): Hmm. I see the sum of the squares on AB and BC = the square on AC.

WIFE: Dear dear.

PYTHAGORAS: I'm not surprised, not surprised at all. I've been saying that would come for years.

WIFE: Yes dear.

PYTHAGORAS: Now they'll hav to *do* something about it. More tea please.

There's another thing—the day is coming when they're going to have

Figure 11.3. A few lazy parrallelograms basking on Mount Olympus. Pythagoras stalking them. (Willans and Searle, p. 45.)

to face the fact that a strate line if infinitely protracted goes on for ever.

WIFE: Quite so.

PYTHAGORAS: Now take the angle a, for xsample.

(*His wife sudenly looses control and thro the porridge at him. Enter Euclid: another weed and the 2 bores go off together*)

Bao Suyo

Kim Stanley Robinson, 1996

It seems appropriate to end this anthology with a passage set in the distant future. Kim Stanley Robinson's *Mars* trilogy envisages the colonization of Mars over a period of more than two centuries, beginning in the 2020s, and the passage reproduced here about Bao Suyo, "the first queen of physics," takes place in the twenty-third century. (Sax Russell, the other character in this extract, is one of the—now extremely elderly—original Martian colonists.) If it bears a relationship with the extracts from science fiction of the seventeenth (*The Blazing-World*) and eighteenth (*Automathes*) centuries we have seen, it also draws something from the other discussions and representations of mathematics and mathematicians—Kanigel's Ramanujan, for instance—that make up this book and brings together some of the particular concerns of the twentieth century: the odd beauty of mathematics and the abiding fear that it may turn out to be a "house of cards."

Kim Stanley Robinson (b. 1952), *Blue Mars* (London, 1996), pp. 432–438. Reprinted by permission of HarperCollins Publishers Ltd. © 1996 Kim Stanley Robinson.

Interesting in a different way was the fact that one of the leading theorists in this new stage of development was working right there in Da Vinci, part of the impressive group Sax was sitting in on. Her name was Bao Suyo. She had been born and raised in Dorsa Brevia, her ancestry Japanese and Polynesian. She was small for one of the young natives, though still half a metre taller than Sax. Black hair, dark skin, Pacific features, very regular and somewhat plain. She was shy with Sax, shy with everyone; she even sometimes stuttered, which Sax found extremely endearing. But when she stood up in the seminar room to give a presentation, she became quite firm in hand if not in voice, writing her equations and notes on the screen very quickly, as if doing speed calligraphy. Everyone in these moments attended to her very closely, in effect mesmerized; she had been working at Da Vinci for a year now, and everyone there smart enough to recognize such a thing knew that they were watching one of the pantheon at work, discovering reality right there before their eyes.

The other young turks would interrupt her to ask questions, of course—there were many good minds in that group—and if they were lucky, off they all would all go together, mathematically modelling gravitons and gravitinos, dark matter and shadow matter—all personality and indeed all persons forgotten. Very productive, exciting sessions; and clearly Bao was the driving force in them, the one they relied on, the one they had to reckon with.

It was a bit disconcerting. Sax had met women in math and physics departments before, but this was the only female mathematical genius he had ever even heard of, in all the long history of mathematical advancement, which, now that he thought of it, had been a weirdly male affair. Was there anything in life as male as mathematics had been? And why was that?

Disconcerting in a different way was the fact that areas of Bao's work were based on the unpublished papers of a Thai mathematician of the previous century, an unstable young man named Samui, who had lived in Bangkok brothels and committed suicide at the age of twenty-three, leaving behind several "last problems" in the manner of Fermat, and insisting to the end that all of his math had been dictated to him by telepathic aliens. Bao had ignored all that and explained some of Samui's more obscure innovations, and then used them to develop a group of expressions called advanced Rovelli-Smolin operators, which allowed her to establish a system of spin networks that meshed with superstrings very beautifully. In effect this was the complete uniting of quantum mechanics and gravity at last, the great problem solved—if it were true. And true or not, it had been powerful enough to allow Bao to make several specific predictions in the larger realms of the atom and the cosmos; and some of these had since been confirmed.

So now she was the queen of physics—the first queen of physics— and experimentalists in labs all over were online to Da Vinci, anxious to have more suggestions from her. The afternoon sessions in the seminar room were invested with a palpable sense of tension and excitement; Max Schnell would start the meeting, and at some point call on Bao; and she would stand and go to the screen at the front of the room, plain, graceful, demure, firm, pen flying over the screen as she gave them a way to calculate precisely the neutrino mass, or described very specifically the ways strings vibrated to form the different quarks, or quantized space so that gravitinos were divided into three families, and so on; and her colleagues and friends, perhaps twenty men and one other woman, would interrupt to ask questions, or add equations that explained side-issues, or tell the rest of them about the latest results from Geneva or Palo Alto or Rutherford; and during that hour, they all knew they were at the centre of the world.

And in labs on Earth and Mars and in the asteroid belt, following her work, unusual gravity waves were noted, in very difficult, delicate experiments; particular geometric patterns were revealed in the fine fluctuations in the cosmic background radiations; dark matter WIMPs and shadow matter WISPs were being sought out; the various families of leptons and fermions and leptoquarks were explained; galactic clumping in the first inflation was provisionally solved; and so on. It seemed as if physics might be on the brink of the Final Theory at last. Or at least in the midst of the Next Big Step.

Given the significance of what Bao was doing, Sax felt shy about speaking to her. He did not want to waste her time on trivial things. But one afternoon at a kava party, out on one of the arc balconies overlooking Da Vinci's crater lake, she approached him—even more shy and stumbling than he was—so much so that he was forced into the very unusual position of trying to put someone else at ease, finishing sentences for her and the like. He did that as best he could, and they stumbled along, talking about his old Russell diagrams for gravitinos, useless now he would have thought, though she said they still helped her to see gravitational action. And then when he asked a question about that day's seminar, she was much more relaxed. Yes, clearly that was the way to put her at ease; he should have thought of it immediately. It was what he liked himself.

After that, they got in the habit of talking from time to time. He always had to work to draw her out, but it was interesting work. And when the dry season came, in the autumn helionequinox, and he started going out sailing again from the little harbour Alpha, he asked her haltingly if she would like to join him, and they stuttered their way through a deeply awkward interaction, which resulted in her going out with him the next nice day, sailing in one of the lab's many little catamarans.

When day sailing Sax stayed in the little bay called the Florentine, southeast of the peninsula, where Ravi Fjord widened but before it became Hydroates Bay. This was where Sax had learned to sail, and where he still felt best acquainted with the winds and currents. On longer trips he had explored the delta of fjords and bays at the bottom end of the Marineris system, and three or four times he had sailed up the eastern side of the large Chryse Gulf, all the way to Mawrth Fjord and along the Sinai Peninsula.

On this special day, however, he confined himself to the Florentine. The wind was from the south, and Sax tacked down into it, enlisting Bao's help at every change of tack. Neither of them said much. Finally, to get things started, Sax was forced to ask about physics. They talked about the ways in which strings constituted the very fabric of spacetime itself, rather than being replacements for points in some absolute abstract grid.

Thinking it over, Sax said, "Do you ever worry that work on a realm so far beyond the reach of experiment will turn out to be a kind of house of cards—knocked over by some simple discrepancy in the math, or some later, different theory that does the job better, or is more confirmable?"

"No," Bao said. "Something so beautiful as this has to be true."

"Hmmm," Sax said, glancing at her. "I must admit I'd rather have something solid crop up. Something like Einstein's Mercury—a known discrepancy in the previous theory, which the new theory resolves."

"Some people would say that the missing shadow matter fits that bill."

"Possibly."

She laughed. "You need more, I can see. Perhaps some kind of thing we can do."

"Not necessarily," Sax said. "Although it would be nice, of course. Convincing, I mean. If something were better understood, so that we could manipulate it better. Like the plasmas in fusion reactors." This was an ongoing problem in another lab at Da Vinci.

"Plasmas might very well be better understood if you modelled them as having patterns imposed by spin networks."

"Really?"

"I think so."

She closed her eyes—as if she could see it all written down, on the inside of her eyelids. Everything in the world. Sax felt a piercing stab of envy, of—loss. He had always wanted that kind of insight; and there it was, right in the boat beside him. Genius was a strange thing to witness.

And they went back to talking about the new results from CERN; about weather; about the sailboat's ability to point to within a few degrees of the wind. And then the following week she went out with him again, on one of his walks on the peninsula's seacliffs. It was a great pleasure to show her a bit of the tundra. And over time, taking him through it step by step, she managed to convince him that they were perhaps coming close to understanding what was happening at the Planck level. A truly amazing thing, he thought, to intuit this level, and then make the speculations and deductions necessary to flesh it out and understand it, creating a very complex powerful physics, for a realm that was so very small, so very far beyond the senses. Awe-inspiring, really. The fabric of reality. Although both of them agreed that just as with all earlier theories, many fundamental questions were left unanswered. It was inevitable. So that they could lie side by side in the grass in the sun, staring as deeply into the petals of a tundra flower as ever one could, and no matter what was happening at the Planck level, in the here and now the petals glowed blue in the light with a quite mysterious power to catch the eye.

Index